INTERMEDIATE

2

physics

Campbell White

handbook

second edition

Hamilton Publishing

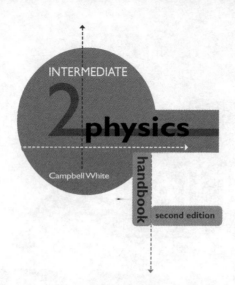

INTERMEDIATE
2 physics
Campbell White
handbook
second edition

First published 1999
Reprinted 2000
Second Edition 2003
© Campbell White 1999, 2003

ISBN 0 946164 53 3 2nd Edition
(ISBN 0 946164 39 8 1st Edition)

*A catalogue record for this book is available from
the British Library.*

**Visit Hamilton Publishing *direct* at
http://www.hamilton-publishing.com**

Orders can be made *direct* over the phone
Contact Thomson Litho, Hamilton Publishing (Sales)
on (01355) 233081

Access and Visa Cards accepted

Letter accepted with school or personal cheque

Published by
Hamilton Publishing
A division of M & A Thomson Litho Limited
10-16 Colvilles Place, Kelvin Industrial Estate,
EAST KILBRIDE G75 0SN

Designed by Smith & Paul

Printed and bound in Great Britain by
M & A Thomson Litho Ltd, East Kilbride, Scotland

Dynamics

Force

A force can change:

- the shape of an object;
- the speed of an object;
- the direction of travel of an object.

The symbol for force is *F*. The unit that is used to measure force is the newton (N).

The newton balance

A **newton balance** or a **spring balance** contains a spring which extends when a force is applied to it. Since the extension is proportional to the force applied to it, a newton balance can be used to measure force.

Weight, mass and the force of gravity

The **weight** of an object is the pull of the Earth on the object. Weight is a force so it is measured in newtons (N).

The **mass** of an object is the quantity of matter that the object contains. The symbol for mass is *m*. The unit that is used to measure mass is the kilogram (kg).

The force that acts on an object because of its mass is the **force of gravity** or the weight of the object.

The force of gravity near the Earth's surface gives all objects the same acceleration (if the effects of friction can be ignored). The approximate value of the acceleration due to gravity near the Earth's surface is 10 m/s^2.

Gravitational field strength

The **gravitational field strength** is the ratio of weight to mass, or the weight per unit mass.
The symbol for gravitational field strength is *g*. The unit that is used to measure gravitational field strength is the newton per kilogram (N/kg). This unit is also equivalent to the metre per second squared (m/s^2), as shown on page 15.

The equation linking mass and weight is:

$$\text{gravitational field strength} = \frac{\text{weight}}{\text{mass}}$$

$$g = \frac{W}{m} \qquad\qquad W = m\,g$$

where   *g* is gravitational field strength in N/kg
       W is weight in N
       m is mass in kg.

The ratio of weight to mass for an object near the Earth's surface is approximately 10 N/kg.

EXAMPLE

Calculate the weight of a person of mass 60 kg.

SOLUTION

mass *m* = 60 kg
gravitational field strength *g* = 10 N/kg (not stated explicitly)
weight *W* = ?
    W = *m g* = 60 x 10 = 600
weight of person = 600 N

Gravitational field strength on other bodies

The gravitational field strength on the Moon and on other planets has different values.

The approximate value of the gravitational field strength on some bodies in the Solar System is as follows.

Heavenly body	Gravitational field strength on the surface in N/kg
Earth	10
Jupiter	26
Mars	4
Mercury	4
Moon	1.6
Neptune	12
Saturn	11
Sun	270
Venus	9

Mass is the amount of 'stuff' or matter that an object contains. Weight is the pull of the Earth or other body on this matter. For any object, the mass remains constant but the weight can vary since it depends on the gravitational field strength.

EXAMPLE

The Lunar Rover, which was used on some Apollo missions, has a mass of 218 kg.
Calculate its weight on the Earth and on the Moon.

The Lunar Rover

SOLUTION

mass m = 218 kg
weight w = ?
gravitational field strength g:
g_{Earth} = 10 N/kg
g_{Moon} = 1.6 N/kg

$W = m\,g$
on Earth $W = 218 \times 10 = 2180$
on Moon $W = 218 \times 1.6 = 348.8$

weight on Earth = 2180 N
weight on Moon = 348.8 N

Friction

Friction is a force which can oppose the motion of an object.

In the following situations, the force of friction is **useful** and should be kept as high as possible.

(i) When **car brakes** are applied to slow a car down, the application of brakes increases the force of friction at the wheels of the car by causing the brake linings to press on the brake drums or discs. This causes the kinetic energy of the car to be converted into heat in the brakes.

The same applies to **bicycle brakes**. Bicycle brakes are more effective when the brake blocks are dry because wet brakes reduce the force of friction.

(ii) The **grips of rackets and bats** used in sports are made of rubber-like material to increase the force of friction between the handle and the hand. This allows a better grip to be maintained.

(iii) In a similar way, **car tyres** keep a good grip on the road because of the friction present.

In the following situations the force of friction is **unwanted** and must be reduced as much as possible.

(i) Friction between moving parts in a **car engine** reduces its efficiency. The force of friction is reduced by lubricating the moving parts with oil. The oil separates the moving parts.

(ii) Vehicles are **streamlined**, making their shape more aerodynamic, to reduce the force of friction due to air resistance. This enables them to move faster.

With a **hovercraft** the friction is reduced even more by lifting it on a cushion of air as it travels over the water.

(iii) When **skiing**, the low value of the force of friction between snow or ice and the ski is reduced even more by waxing the ski surface to make it as smooth as possible.

Streamlined car

Balanced forces

Force is a vector quantity and so needs a direction as well as a size to describe it fully.

Equal forces acting in opposite directions on an object are called **balanced forces** and are equivalent to no force at all.

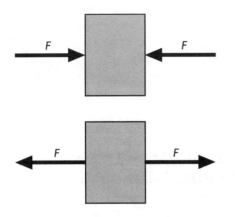

If two or more forces acting on an object balance overall, they have the same effect as no force at all. For example:

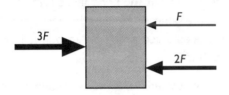

Seat belts are used in a car to supply a restraining force to stop the continued forward motion of the occupants when the car stops suddenly. If this does happen, an unrestrained occupant would continue to move forward at the same original velocity and would probably hit the windscreen. A seat belt provides a backward force to reduce this velocity to that of the reduced velocity of the car.

When an object is dropped from a height, its velocity increases (it accelerates) as it drops towards the surface of the Earth. A parachute is used to increase the air resistance of a falling object so that its acceleration is reduced. When the upwards force of the air resistance is equal to the downwards weight of the object then the forces are balanced and the velocity of the object is constant. It is then falling at its **terminal velocity**.

Newton's First Law of Motion
Newton's First Law of Motion states:

> An object will continue in its state of rest or of uniform motion unless it is acted upon by an external, unbalanced force.

When an object has no forces or balanced forces acting on it, its velocity stays the same.

An object sitting on a surface does not move because the forces acting on it are balanced.

A vehicle can travel along a straight level road at a constant speed if there are no unbalanced forces acting on it. For this to happen the force of friction which is present when the vehicle moves must be balanced by the force supplied by the engine.

A spacecraft in outer space with no engines on continues moving at the same velocity because there are no forces acting on it, so once it is moving it will continue with the same velocity.

Resultant force and free body diagrams

If the forces acting on an object do not balance each other, then the extra force is called the **resultant force**. The combined effect of two or more forces acting on an object is the resultant of the forces and is the vector sum of those forces.

A **free body diagram** shows the forces acting at one point in a system or more simply can be used to find the resultant of forces acting on an object. A free body diagram is so named because it is a vector diagram of the forces acting at one suitable point and does not include the object (body).

EXAMPLE

A stone is pulled back in a catapult until the elastic in each side of the catapult makes an angle of 30° with the direction of pull. Draw a free body diagram showing the horizontal forces acting on the stone.

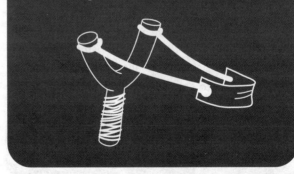

SOLUTION

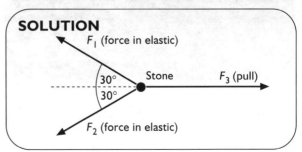

F_1 (force in elastic), F_3 (pull), Stone, 30°, 30°, F_2 (force in elastic)

The resultant of two forces can be found either by drawing a scale diagram or by using trigonometry. Whichever method is used, since the resultant is a force which is a vector quantity, both the size and the direction must be found.

To find the resultant by drawing a scale diagram, use the following procedure.

(i) Choose and state a suitable scale.

(ii) Draw a vector (a directed line) to scale, to represent one of the forces.

(iii) From the head of this vector draw a second vector to represent the second force.

(iv) The resultant of the two forces is represented by the vector obtained by joining the tail of the first vector to the head of the second vector.

(v) State the size of the resultant by measuring the length of this vector and using the scale in reverse. State the direction by measuring a suitable angle and quoting it in an appropriate way, for example $x°$ east of north.

To find the resultant by trigonometry, use the following procedure.

(i) Sketch a vector diagram for the forces given, not necessarily to a scale. Include the resultant on this diagram.

(ii) Use Pythagoras' relationship for the triangle formed to find the magnitude (size) of the resultant.

(iii) Use the trigonometrical relationships of sin, cos or tan as appropriate to calculate a suitable angle in this triangle.

(iv) State the resultant by giving the size and quoting the angle calculated in an appropriate way, for example $x°$ to the bank of the river.

EXAMPLE

A force of 4N to the east is applied to an object at the same time as a force of 3N to the south.
Use a scale diagram to find the magnitude and the direction of the resultant of the two forces.

SOLUTION

Scale: 1 cm represents 1 N

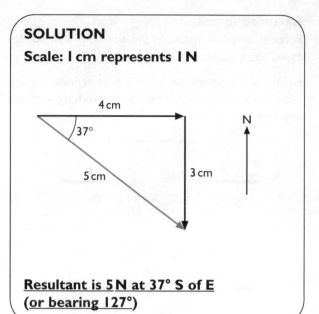

Resultant is 5 N at 37° S of E (or bearing 127°)

Newton's Second Law of Motion

An **unbalanced** or **resultant force** causes an object to have an acceleration. When the mass of an object decreases but the unbalanced force acting on it remains the same, its acceleration increases. When the unbalanced force acting on an object increases, its acceleration increases.

The **newton**, the unit of force, is defined in terms of the acceleration produced in an object by an unbalanced or resultant force. The newton is the force which causes an object of mass 1 kg to have an acceleration of 1 m/s².

The relationship between acceleration, resultant or unbalanced force and mass is:

$$\text{acceleration} = \frac{\text{unbalanced force}}{\text{mass}}$$

$$a = \frac{F}{m}$$

where a is acceleration in m/s²
      F is unbalanced force in N
      m is mass in kg.

This is a version of **Newton's Second Law of Motion**, which says:

The acceleration produced in an object is directly proportional to the unbalanced force acting on it and inversely proportional to its mass.

EXAMPLE

A car of mass 1000 kg is being driven along a straight, level road. The engine supplies a driving force of 2500 N and the total resistive forces due to friction and air resistance amount to 1000 N.

(a)   Draw a free body diagram showing the forces acting on the car.
(b)   Calculate its acceleration.

SOLUTION

(a)   mass m = 1000 kg
       driving force = 2500 N
       resistive forces = 1000 N
       so unbalanced force,
       $F = (2500 - 1000)N = 1500 N$

Resistive forces   Driving force

1000 N   2500 N

Free body diagram

(b)
$$\text{acceleration} = \frac{\text{unbalanced force}}{\text{mass}}$$

$$a = \frac{F}{m}$$

$$a = \frac{1500}{1000} = 1.5$$

acceleration = 1.5 m/s²

The equivalence of acceleration due to gravity and gravitational field strength

The acceleration due to gravity is equivalent to gravitational field strength as can be seen from the following.

Gravitational field strength g is weight per unit mass.

$$g = \frac{W}{m}$$

An object free to move in a gravitational field will accelerate because of the resultant (unbalanced) force acting on it, called its weight, obeying Newton's Second Law.

$$W = m\,a \qquad (a \text{ is the acceleration due to gravity})$$

Combining these equations gives:

$$g = \frac{W}{m} = \frac{m\,a}{m} = a$$

So gravitational field strength = acceleration due to gravity.
The units of both quantities, N/kg and m/s², are also equivalent.

Projectiles

If an object is projected horizontally it follows a curved (parabolic) path. This is because the force of gravity accelerates it in the downward (vertical) direction, while it travels forwards (horizontally) at a constant speed.

The horizontal motion of a projectile shows constant speed because no forces act on the projectile horizontally (ignoring air resistance). The horizontal motion of a projectile is governed by Newton's First Law.

The vertical motion of a projectile shows uniform acceleration, caused by gravity. So the vertical acceleration of a projectile near the Earth's surface is 10 m/s². The vertical motion of a projectile is governed by Newton's Second Law.

Projectile motion can be treated as two independent motions where the only quantity that is common to the vertical and horizontal motions of a projectile is the **time of flight.**

EXAMPLE

A coin is rolled off the edge of a table at a speed of 1 m/s and falls to the floor, taking a time of 0.5 s.

(a) What happens to the horizontal speed of the coin while it is falling?

(b) What happens to the vertical speed of the coin while it is falling?

(c) Calculate how far out from the edge of the table the coin lands.

SOLUTION

(a) **Horizontal speed remains constant at 1 m/s.**

(b) **Vertical speed increases uniformly from 0 m/s.**

(c) **Horizontal motion:**

$v = 1$ m/s
$t = 0.5$ s
$s = ?$
$s = vt$
$\quad = 1 \times 0.5$
$\quad = 0.5$

The coin lands 0.5 m from the edge of the table.

Solving projectile problems can involve using a velocity–time graph to find vertical displacement.

EXAMPLE

An archer shoots an arrow horizontally at a target 50 m away. The horizontal speed of the arrow is 100 m/s. How much below the centre of the target does the arrow hit, if it was aimed directly at the centre?

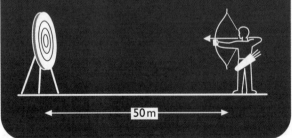

50 m

SOLUTION

Horizontal motion:
velocity $v = 100$ m/s
distance $s = 50$ m
time $t = ?$

$$t = \frac{s}{v} = \frac{50}{100} = 0.5$$

Time taken for arrow to hit target = 0.5 s

Vertical motion:
initial velocity $u = 0$
acceleration $a = 10$ m/s²
time $t = 0.5$ s
final velocity $v = ?$
$v = u + at = 0 + 10 \times 0.5 = 5$
Final velocity = 5 m/s

Plot a velocity-time graph for this vertical motion.

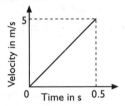

Vertical displacement (the distance the arrow falls) is the area under the graph
displacement $= \frac{1}{2} \times 0.5 \times 5 = 1.25$
The arrow will hit 1.25 m below the centre of the target.

Newton's thought experiment and satellite motion

Long before artificial satellites were put into orbit, Sir Isaac Newton used projectile motion to explain the orbit of satellites in his famous thought experiment. He suggested that a large cannon, fired horizontally from the top of a tall mountain, would have a very large range – well beyond the horizon. He extended the idea to an even taller mountain and an even bigger cannon and suggested that, if it were to fire a shell with a great enough velocity, the shell would always fall towards the Earth but would never reach it. In effect the shell would orbit the Earth because of gravity and would become an artificial satellite.

A **satellite**, like a projectile, is constantly falling towards the Earth. However, the forward speed of a satellite is so great that because of the curvature of the Earth, the Earth's surface drops away as much as the satellite falls towards it. So the satellite always stays at the same height above the Earth's surface.

The same applies to all satellites, natural or man-made, which orbit any planet or heavenly body.

Momentum and energy

Newton's Third Law of Motion

Newton's Third Law states:

> To every action force there is an equal and opposite reaction force.

This law is sometimes quoted as 'action and reaction are equal but opposite' or as 'if A exerts a force on B, B exerts an equal but opposite force on A'.

'Newton Pairs' are the two equal but opposite forces which act on each of the two objects A and B which are mentioned in one version of the statement of Newton's Third Law.

EXAMPLE

Identify the 'Newton Pairs' acting on the helicopter shown.

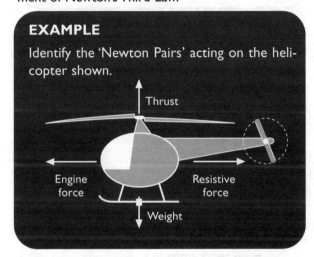

SOLUTION

The 'Newton Pairs' acting on the helicopter are as follows.

Thrust: the force of the rotors (A) on the air (B) and the force of the air (B) on the rotors (A)

Weight: the force of the Earth (A) on the helicopter (B) and the force of the helicopter (B) on the Earth (A)

Engine force: the force of the engines (A) on the air (B) and the force of the air (B) on the engines (A)

Resistive force: the force of friction of the air (A) on the helicopter (B) and the force of the helicopter (B) on the air (A)

Momentum

When two cars collide, or vehicles are involved in a motorway pile-up, investigations afterwards show that the damage caused depends on two factors: the masses of the vehicles involved and their velocities. From these findings, it is apparent that the product of mass and velocity has a particular importance when collisions are concerned.

The product of mass and velocity is called **momentum**.

> momentum = mass x velocity

Momentum is not often given a symbol, although sometimes p is used.

The unit that is used to measure momentum is the kilogram metre per second (kg m/s) – there is no simpler unit for this quantity.

Since momentum involves velocity which is a vector quantity, momentum is also a vector quantity and so requires a direction as well as a size to describe it fully.

Conservation of momentum

The **Law of Conservation of Linear Momentum** as it relates to two objects involved in a collision (or other interaction) in one direction is:

total momentum before collision = total momentum after collision

This relationship is often written as an equation:

$$m_1 u_1 + m_2 u_2 = m_1 v_1 + m_2 v_2$$

where m is mass in kg

u is velocity before the collision in m/s

v is velocity after the collision in m/s

the subscript 1 is used for object 1 in the collision

the subscript 2 is used for object 2 in the collision.

EXAMPLE

During an experiment, a trolley of mass 0.75 kg travelling at 1 m/s collides with and sticks to a second, identical trolley which is initially at rest.
Calculate the velocity at which the combined trolleys move off.

SOLUTION

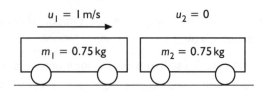

Before collision

$v_1 = v_2 = ?$

After collision

total momentum before collision
= total momentum after collision
$$m_1 u_1 + m_2 u_2 = m_1 v_1 + m_2 v_2$$
$$(0.75 \times 1) + (0.75 \times 0) = (0.75 + 0.75) \times v$$
$$0.75 = 1.5 \, v$$
$$v = 0.5$$
velocity of trolleys after collision
= 0.5 m/s

Work and energy

When work is done by a force, **energy** is transferred and the amount of work that is done is a measure of the **energy transferred**. The symbol for all types of energy is E. The two symbols that are used to denote **work done** are W and E_w. The unit used to measure both energy and work done is the joule (J).

The equation to calculate work done is:

$$\text{work done} = \text{force} \times \text{distance}$$
$$W = F \, s$$

where W is work done in J
    F is force in N
    s is distance in m.

Although the work done by a force is a scalar quantity, the object acted upon moves in the direction of the applied force.

EXAMPLE

A force of 80 N is used to move a crate a distance of 3 m along the floor.
Calculate the amount of work done on the crate.

SOLUTION

force $F = 80 \, N$
distance $s = 3 \, m$
work done $W = ?$
    $W = F s = 80 \times 3 = 240$
work done on crate = 240 J

Power

Power is the rate at which work is done or energy is transferred. The symbol for power is P. The unit that is used to measure power is the watt (W). The relationship between power, work done and time is:

$$\text{power} = \frac{\text{work done}}{\text{time}}$$

$$P = \frac{E_w}{t}$$

where P is power in W
    E_w is work done in J
    t is time in s.

EXAMPLE

A toy car of mass 0.5 kg takes 8 s to move along a slope, rising a height of 2 m in doing so.
(a)   Calculate the weight of the car.
(b)   Calculate the amount of work done (energy transferred) by the car.
(c)   Calculate the power developed by the car in moving up the slope.

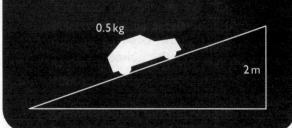

SOLUTION

(a) mass m = 0.5 kg
  gravitational field strength g = 10 N/kg (not stated explicitly)
  weight = ?
    weight = $m g$ = 0.5 x 10 = 5
    <u>weight of toy car = 5 N</u>

(b) force F = weight = 5 N
  distance s = height = 2 m
  work done = ?
    work done = $F s$ = 5 x 2 = 10
    <u>work done by toy car = 10 J</u>

(c) work done E_w = 10 J
  time t = 8 s
  power P = ?

  $$P = \frac{E_w}{t} = \frac{10}{8} = 1.25$$

<u>power developed by toy car = 1.25 W</u>

SOLUTION

mass m = 60 kg
gravitational field strength g = 10 N/kg (not stated explicitly)
height h = 80 m
gravitational potential energy E_p = ?
  $E_p = m g h$ = 60 x 10 x 80 = 48 000
<u>increase in gravitational potential energy = 48 000 J</u>

Potential energy

When an object is lifted up in a gravitational field, work has to be done on the object against gravity and this work is stored in the object as an increase in **gravitational potential energy**.

If an object is moved to a lower position in a gravitational field, then the energy stored as gravitational potential energy can be regained. Here work is done *by* the gravitational field.

The symbol for gravitational potential energy is E_p. The unit that is used to measure gravitational potential energy is the joule (J).

The equation to calculate gravitational potential energy is:

  $E_p = m g h$

where E_p is gravitational potential energy in J
  m is mass in kg
  g is gravitational field strength in N/kg
  (or acceleration due to gravity in m/s²)
  h is change in height in m.

EXAMPLE

A student of mass 60 kg climbs 80 m up a hill. Calculate the increase in gravitational potential energy of the student.

Kinetic energy

An object has **kinetic energy** because of its motion.

The greater the **mass** and/or the greater the **velocity** of a moving object, the greater its kinetic energy is.

The symbol for kinetic energy is E_k. The unit that is used to measure kinetic energy is the joule (J).

The equation to calculate kinetic energy is:

  $E_k = \frac{1}{2} m v^2$

where E_k is kinetic energy in J
  m is mass in kg
  v is speed in m/s.

From this relationship, it can be seen that the amount of kinetic energy that an object has is proportional to the *square* of its velocity. This means that, if a driver doubles a car's velocity, four times the amount of kinetic energy will need to be transformed to bring the car to rest. Since a driver also takes time to think about stopping, even in an emergency (because of reaction time), a fast-moving vehicle could travel a considerable distance before coming to rest.

EXAMPLE

Calculate the kinetic energy of a car of mass 1000 kg when it is travelling with a velocity of 12 m/s.

SOLUTION

mass m = 1000 kg
velocity v = 12 m/s
kinetic energy E_k = ?
  $E_k = \frac{1}{2} m v^2 = \frac{1}{2}$ x 1000 x 12² = 72 000
<u>kinetic energy of car = 72 000 J</u>

Machines and efficiency

A **machine** is a device which is used to transform or change energy from one form to another. In any energy transformation, as well as a useful form of energy being produced, some energy is also transformed into useless forms. The least organised and most uncontrollable form of energy is heat. It is often 'lost' to the atmosphere. This is referred to as energy being **degraded** during an energy transformation.

The **efficiency** of an energy transformation is a way of expressing how much useful energy is obtained for a given energy input. Efficiency is often expressed as a percentage.

One equation to calculate efficiency is:

$$\text{efficiency} = \frac{\text{useful energy out}}{\text{total energy in}}$$

$$\text{efficiency} = \frac{E_{out}}{E_{in}}$$

$$\text{percentage efficiency} = \frac{E_{out}}{E_{in}} \times 100$$

E_{out} and E_{in} can be in any suitable units, as long as they are the same.
Efficiency is a ratio and so does not have any units.

Since the input and output *times* involved in any energy transformation must be the same, input power and output power can be used to calculate efficiency:

$$\text{efficiency} = \frac{\text{useful power out}}{\text{total power in}}$$

$$\text{efficiency} = \frac{P_{out}}{P_{in}}$$

$$\text{percentage efficiency} = \frac{P_{out}}{P_{in}} \times 100$$

P_{out} and P_{in} can be in any suitable units, as long as they are the same.
Efficiency does not have any units.

EXAMPLE

A microwave oven has an input power of 800 W and an output power of 650 W. Calculate its percentage efficiency.

SOLUTION

input power P_{in} = 800 W
output power P_{out} = 650 W
percentage efficiency = ?

$$\text{percentage efficiency} = \frac{P_{out}}{P_{in}} \times 100$$

$$= \frac{650}{800} \times 100 = 81.25$$

efficiency of oven = 81.25%

Heat

Heat and temperature

Temperature is a measure of how hot or cold an object is. The most common scale used to measure temperature is the **Celsius** scale. It is sometimes called the **centigrade** scale because there are 100 equal divisions between its two fixed points – the melting point of pure melting ice (0°C) and the boiling point of pure water (100°C).

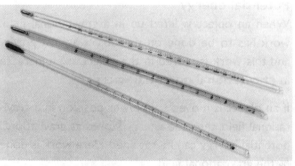

Heat is a form of energy and is measured in joules (J). Putting heat into an object usually but not always makes its temperature increase. Heat travels from a region of high temperature to a region of lower temperature.

The relationship between heat and the temperature change produced in a material can be understood by considering the following.

- One kilogram of copper does not need the same amount of energy as one kilogram of aluminium to raise its temperature by 1°C.

- Two kilograms of copper need twice as much energy to raise the temperature by 1°C as one kilogram does.

- It takes twice as much energy to raise the temperature of one kilogram of copper by 2°C as it does to raise the temperature by 1°C.

- It takes the same amount of energy to raise the temperature of one kilogram of copper from 9°C to 10°C as it does to raise the temperature from 99°C to 100°C.

Specific heat capacity

The **specific heat capacity** of a substance is the amount of energy in joules needed to change the temperature of 1 kg of the substance by 1°C. The value is different for different substances. It is called *specific* heat capacity because the mass of the substance has been specified as 1 kg. Tables are available giving the specific heat capacities of various common substances (see data sheet on page 6).

The symbol for specific heat capacity is *c*. The unit that is used to measure specific heat capacity is the joule per kilogram degree Celsius (J/kg°C).

The equation linking heat and change in temperature to the specific heat capacity and the mass of a substance is:

$$E_h = c\, m\, \Delta T$$

where E_h is heat in J
c is specific heat capacity in J/kg°C
m is mass in kg
ΔT is change in temperature in °C.

Different materials need different amounts of heat to produce a certain change in temperature. Materials that take a lot of heat to produce a small increase in temperature have high heat capacities. Such materials can store a lot of heat at relatively low temperatures, heat which can be released slowly from the material as it cools down. Materials like these are used in storage heaters.

EXAMPLE

Calculate the amount of heat needed to increase the temperature of 2 kg of copper by 150°C.
(The specific heat capacity of copper is 386 J/kg°C.)

SOLUTION

mass *m* = 2 kg
temperature rise Δ*T* = 150°C
specific heat capacity of
copper *c* = 386 J/kg°C
heat E_h = ?
   $E_h = c\, m\, \Delta T = 386 \times 2 \times 150 = 115\,800$
heat needed = 115 800 J

Change of state

The three states of matter are **solid, liquid** and **gas**.

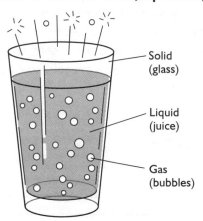

The three states of matter

When heat is transferred to a substance it can increase the temperature of the substance or it can change the state of the substance. If heat is transferred to a substance which is at its melting point or its boiling point already, its state will change. Otherwise, its temperature will increase.

The temperature does not change when the state of a substance is changed.

Energy, usually in the form of heat, must be added to a substance to make its state change from a solid to a liquid or from a liquid to a gas.

Energy, usually in the form of heat, is lost from a substance when its state changes from a gas to a liquid or from a liquid to a solid.

A **cooling curve** is a graph showing how the temperature of a substance changes as time passes while the substance is giving out heat.

When water cools from 100°C to room temperature (about 20°C), it does not undergo a change of state. Since the rate of heat loss is proportional to the temperature difference, its cooling curve is as shown.

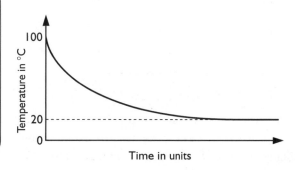

The cooling curve for a substance which changes state at some period over the temperature range observed will have a horizontal section, showing that there is no change in temperature while the state is changing. For example the cooling curve for a substance which changes from solid to liquid at 80°C is as shown.

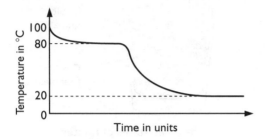

The picnic box cooler

Applications of change of state

A liquid takes in heat when it evaporates, so it leaves its surroundings cooler. This is observed when a liquid such as methanol is allowed to evaporate on the hand.

A similar effect happens when a solid melts. Doctors and physiotherapists use ice packs or similar packs containing chemicals which change from solid to liquid to cool down muscle injuries.

Two examples of applications which involve a change of state are the **refrigerator** and the **picnic box coolant**.

The refrigerator

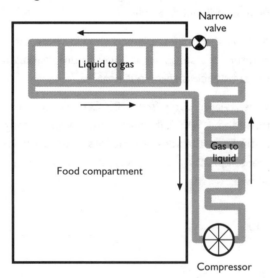

In a refrigerator a liquid with a low boiling point (originally freon but now replaced by ozone-friendly chemicals) is forced through a narrow valve. As it does so, it is changed into a gas, taking the heat to do

so from the contents of the food compartment inside the refrigerator so cooling them down. When it moves outside the cabinet, it is passed through a compressor where it is converted back into a liquid.

A picnic box cooler is a sealed plastic pack which contains a chemical that can readily be solidified by placing it for some time in a refrigerator. When placed near the top of a cool box, the chemical gradually converts into a liquid again by taking heat from the cool box contents, so keeping them cool.

Latent heat

The term **latent heat** describes heat that cannot be observed or recorded on a thermometer as a change in temperature because latent heat is used to change the state of a substance and a change of state does not involve a change in temperature.

The **latent heat of fusion** of a substance is the amount of heat needed to change unit mass of the substance from a solid to a liquid, or the heat given out when unit mass of the substance changes from a liquid to a solid. The value is different for different substances. Tables are available giving the latent heat of fusion of various common substances (see the data sheet on page 6).

The **latent heat of vaporisation** of a substance is the amount of heat needed to change unit mass of the substance from a liquid to a gas, or the heat given out when unit mass of the substance changes from a gas to a liquid. The value is different for different substances. Tables are available giving the latent heat of vaporisation of various common substances (see the data sheet on page 6).

The value for the latent heat of vaporisation of a substance is always greater than the value for the latent heat of fusion of the same substance.

The symbol for **latent heat** is l. The unit that is used to measure latent heat is the joule per kilogram (J/kg). Since the mass has been specified as 1 kg this quantity is usually referred to as the **specific latent heat** of a substance.

The equation to calculate **specific latent heat** is:

$$E_h = m\, l$$

where E_h is heat in J

      m is mass in kg

      l is specific latent heat of fusion or vaporisation as appropriate in J/kg.

EXAMPLE

Calculate the amount of heat needed to melt an ice cube of mass 50 g without changing its temperature.

SOLUTION

mass m = 50 g = 0.05 kg

specific latent heat of fusion of ice l = 3.34 x 10⁵ J/kg (not stated explicitly)

heat E_h = ?

   $E_h = m\, l = 0.05 \times 3.34 \times 10^5 = 16\,700$

heat needed = 16 700 J

EXAMPLE

10 g of steam condense on a window. Calculate the amount of heat given out during the condensation process.

SOLUTION

mass m = 10 g = 0.01 kg

specific latent heat of vaporisation of water l = 2.26 x 10⁶ J/kg (not stated explicitly)

heat E_h = ?

   $E_h = m\, l = 0.01 \times 2.26 \times 10^6 = 22\,600$

heat given out = 22 600 J

Conservation of energy

An **energy transformation** happens when energy is changed from one form to another.

The **principle of conservation of energy** says that, although energy can be transformed from one type to another, it can neither be created nor destroyed.

Since energy is conserved during any transformation, if the energy at any point can be calculated, this means that the amount of energy at all points is known. It is usual in problems of this type to ignore the fact that energy is transformed into unwanted heat, unless stated otherwise.

EXAMPLE

A crane lifts a crate of mass 60 kg through a height of 25 m in a time of one minute.

(a) Calculate the gravitational potential energy gained by the crate.

(b) Ignoring the mass of the crane jib, calculate the output power of the crane.

(c) If the cable of the crane breaks when it has completed the lifting operation, calculate the maximum velocity of the crate just as it reaches the ground (ignoring air resistance).

SOLUTION

mass m = 60 kg

height h = 25 m

time t = 1 minute = 60 s

gravitational field strength g = 10 m/s² (not stated explicitly)

(a) $E_p = m\, g\, h = 60 \times 10 \times 25 = 15\,000$

   gravitational potential energy gained = 15 000 J

(b) **Power relationship:**

$$P = \frac{E}{t} = \frac{15\,000}{60} = 250$$

   output power of crane = 250 W

(c) If the crate falls, all of its potential energy will be transformed into kinetic energy.

   $E_k = 15\,000 = \frac{1}{2} m\, v^2$

   $15\,000 = \frac{1}{2} \times 60 \times v^2$

$$v^2 = \frac{15\,000 \times 2}{60} = 500$$

   velocity of crate = 22.36 m/s

MECHANICS AND HEAT QUESTIONS

Kinematics

Average speed

1. Explain what is meant by **average speed**.

page 7

2. State the relationship between distance, time and average speed.

page 7

3. (a) What is the symbol that is used for average speed?

(b) What is the unit, and its abbreviation, that is used to measure average speed?

page 7

4. Describe how to measure the average speed of a runner in a race.

Your description should include:

the measurements that have to be made;
any special points about these measurements;
how these measurements are made;
the equation that is used to calculate the speed.

page 7

5. A runner completes a 200 m race in 25 s. Calculate his average speed.

6. A motorist travels the 72 km from Edinburgh to Glasgow in a time of $1\frac{1}{4}$ hours.

Calculate the average speed for this journey in metres per second.

7. Calculate how far a supersonic aircraft would travel in one minute while flying at the speed of sound (340 m/s).

Instantaneous speed

8. Explain what is meant by **instantaneous speed**.

page 7

9. Describe how to measure the instantaneous speed of an object.

Your description should include:

the measurements that have to be made;
any special points about these measurements;
how these measurements are made;
the equation that is used to calculate the speed.

page 7

10. Two towns are 30 km apart. The bus journey from one town to the other takes half an hour.

(a) Calculate the average speed, in kilometres per hour, for the bus during this journey.

(b) At various times during the journey, the bus stops to pick up passengers, travels in heavy traffic in a town centre and travels on a motorway.

Explain why the instantaneous speed of the bus varies between 0 and 80 km/h.

(c) Why are the average and instantaneous speeds of the bus different?

Vector and scalar quantities

11. (a) Describe what is meant by a **scalar quantity.**

(b) Give an example of a **scalar quantity**.

page 8

12. (a) Describe what is meant by a **vector quantity.**

(b) Give an example of a **vector quantity**.

page 8

13. Copy the two tables below and complete them by entering the following quantities under the correct headings.

speed
velocity
acceleration due to gravity
energy
work done
temperature
heat
time
weight
mass
gravitational field strength

Scalar quantities	Vector quantities

14. State the difference between **distance** and **displacement**. *page 8*

15. State the difference between **speed** and **velocity**. *page 8*

16. Which of the following statements describe the **speed** and which the **velocity** of the item concerned?
    (a) A car moving at 15 m/s.
    (b) A boat sailing up the Forth at 0.5 m/s.
    (c) Light travels at 3×10^8 m/s.
    (d) The electron travelled at 2.5×10^8 m/s in the tube.
    (e) The aircraft cruised at 900 kilometres per hour on a south-westerly course to South America.

Acceleration

17. Explain the terms **speed, velocity** and **acceleration**. *page 8*

18. (a) What is the symbol that is used for acceleration?
    (b) What is the unit, and its abbreviation, that is used to measure acceleration? *page 9*

19. What is the acceleration of a car which is travelling at a steady speed along a straight level road?

20. (a) What does a negative sign associated with an acceleration value signify?
    (b) What is another term used for a negative acceleration?

21. A car advertisement makes the following statement relating to the performance of the car: '0 to 26 m/s in 8.2 seconds'.
    What quantity do these figures for the car refer to?

22. Give the equation which links acceleration, initial velocity, final velocity and the time for the velocity to change, defining all of the symbols used in the equation. *page 9*

23. Starting with the definition of acceleration, show that $v = u + at$, where all of the symbols have the usual meanings.

24. Calculate the acceleration of a car that increases its velocity by 10 m/s in 5 s.

25. A car can accelerate from 0 m/s to 26 m/s in 8 s. Calculate its maximum acceleration.

26. Calculate the final velocity of a train which accelerates uniformly at a rate of 0.6 m/s from a velocity of 2 m/s for 30 s.

27. A vehicle accelerates from rest to a velocity of 6 m/s in 3 s.
    Calculate its acceleration.

28. A ship has a maximum acceleration of 0.1 m/s². Calculate the minimum time it would take to increase its velocity from 1 m/s to 5 m/s.

29. A car, decelerating uniformly at 2 m/s², comes to rest in 10 s.
    Calculate its initial velocity.

Velocity–time graphs

30. Draw the velocity–time graph for the motion of a car which is travelling at a steady velocity of 15 m/s.

31. A car pulls away from rest at traffic lights and reaches a velocity of 10 m/s in 8 s.
    Draw the velocity–time graph for this motion.

32. The driver of a car sees the traffic lights in the distance change to red and brakes uniformly. The velocity of the car reduces from 20 m/s to 5 m/s in 6 s.
    Draw the velocity–time graph for this motion.

33. Describe the motion represented by each of the following velocity–time graphs and for each calculate the acceleration shown.

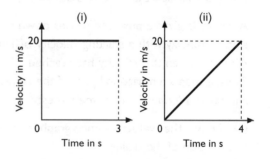

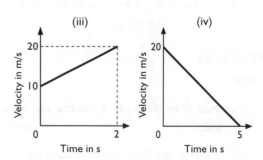

34. How can **displacement** be calculated from a velocity–time graph? *page 9*

35. Consider the following velocity–time graph for the motion of a car:

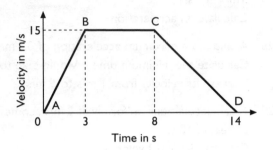

   (a) Describe the motion represented by each section of this graph.
   (b) What is the maximum velocity reached by the car?
   (c) Calculate the acceleration shown on the graph.
   (d) Calculate the deceleration shown on the graph.
   (e) Calculate the displacement of the car after 14 s.

36. A suburban train pulls away from one station with a uniform acceleration and reaches a speed of 25 m/s in 10 s. It then decelerates uniformly to stop at a second station in a *further* 20 s.

   (a) Draw the velocity–time graph for the motion of the train.
   (b) Calculate the acceleration of the train.
   (c) Calculate the deceleration of the train.
   (d) How far apart are the two stations?

37. A trolley is given a push and moves down a sloping runway with a starting velocity of 1 m/s. After 1 s, when the velocity has reached 2 m/s, it meets a horizontal part of the runway and takes a *further* 2 s to come to rest.

   (a) Draw the velocity–time graph for the motion of the trolley.
   (b) Calculate the total length of the runway used by the trolley.

Dynamics

Force

1. Describe *three* effects that a **force** can have on an object. *page 11*

2. (a) What is the symbol that is used for force?
   (b) What is the unit, and its abbreviation, that is used to measure force? *page 11*

The newton balance

3. Describe what a **newton balance** is. A diagram will help your description. *page 11*

4. Give *one* use for a newton balance. *page 11*

Weight, mass and the force of gravity

5. (a) What is meant by the **weight** of an object?
   (b) What is the unit, and its abbreviation, that is used to measure weight? *page 11*

6. Copy and complete:
   Weight is a _____ and so it is measured in _____ . *page 11*

7. What causes an object to have a weight? *page 11*

8. What is meant by the **mass** of an object? *page 11*

9. (a) What is the symbol that is used for mass?
   (b) What is the unit, and its abbreviation, that is used to measure mass? *page 11*

10. Explain clearly the difference between mass and weight. In your explanation, say which of these quantities is a constant and which can vary, and explain why this is so. *page 11*

11. (a) What is the name of the force that acts on an object because of its mass?
    (b) What can be noted about the acceleration of all objects near to the Earth's surface due to this force acting on the objects, if the effects of friction can be ignored? *page 11*

12. State the approximate value of the acceleration due to gravity near the Earth's surface. *page 11*

Gravitational field strength

13. What is meant by **gravitational field strength**? *page 11*

14. (a) What is the symbol that is used for gravitational field strength?
    (b) What is the unit, and its abbreviation, that is used to measure gravitational field strength? *page 11*

15. What is the approximate answer when the weight of an object near the Earth's surface is divided by its mass? *page 11*

16. Calculate the weight on Earth of a person who has a mass of 50 kg.

17. Calculate the weight of a 1 kg bag of sugar.

18. A lorry is weighed on a public weigh bridge and is found to have a weight of 100 000 N. Calculate the mass of the lorry.

Gravitational field strength on other bodies

19. Copy and complete the following statements, using words or phrases from this list:
    stays the same; increases; decreases; becomes zero.
    A block of metal has a mass of 1 kilogram on the Earth.
    (a) When the block is taken to the Moon where the gravitational field strength is 1.6 N/kg its mass _____ and its weight _____ .

    (b) When the block is taken to Jupiter where the gravitational field strength is 26 N/kg its mass _____ and its weight _____ .

    (c) When the block is taken into space, far away from any planets, its mass _____ and its weight _____

20. A piece of equipment which has a weight of 25 N on Earth is taken to the Moon where the gravitational field strength is 1.6 N/kg. Calculate its weight on the Moon.

21. It might be possible, although unlikely, for an astronaut of the future to go 'space-hopping' from planet to planet.
    Calculate the total weight of the spaceman and his spacesuit on each of the planets given in the table, if their combined mass is 120 kg.

Planet	Gravitational field strength (N/kg)
Venus	9
Earth	10
Mars	4
Jupiter	26

Friction

22. Explain what is meant by **friction**.   *page 12*

23. (a) Describe *three* situations where the force of friction is *useful*.
    (b) Explain how the force of friction can be *increased* in these situations.   *page 12*

24. (a) Describe *three* situations where the force of friction is *unwanted*.
    (b) Explain how the force of friction can be *decreased* in these situations.   *page 12*

Balanced forces

25. Which type of quantity is force, a scalar quantity or a vector quantity?   *page 13*

26. Describe what is meant by **balanced forces** acting on an object, using a diagram to help your description.   *page 13*

27. Describe a situation where *two or more* forces can act on an object and yet have the same effect as no force at all.
    Include a diagram in your answer.   *page 13*

28. A book is sitting on a table.
    What can be said about the forces that are acting on the book?

29. A car is travelling along a straight level road at a constant speed.
    (a) What can be said about the forces acting on the car?
    (b) Explain why the car engine needs to be on to maintain the steady speed.

30. Explain why you continually have to pedal a bicycle to move along a straight, level road at a constant speed.

31. (a) What is the purpose of **seat belts** in a car?
    (b) Explain how seat belts in a car do the job that they are designed to do.
    Your explanation should concentrate on the forces involved.   *page 13*

Newton's First Law of Motion

32. State **Newton's First Law of Motion**.   *page 13*

33. What happens to the velocity of an object when it has *no* forces acting on it?   *page 13*

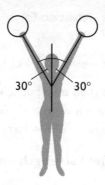

34. What happens to the velocity of an object when it has balanced forces acting on it? *page 13*

35. Use Newton's First Law of Motion to explain the following situations:

    (a) why a book sitting on a table does not move;
    (b) how a car can travel along a straight level road at a constant velocity;
    (c) why a spaceship in outer space continues moving at the same velocity in the same direction. *page 13*

Resultant force and free body diagrams

36. What does an **unbalanced** or **resultant force** cause in an object? *page 13*

37. What happens to an object when its mass decreases if the unbalanced force acting on it remains the same? *page 13*

38. What happens to an object when the unbalanced force acting on it increases? *page 13*

39. A boy on a skateboard is travelling at a constant velocity along a straight, level track. His friend jumps on to his skateboard.
    What happens to the motion of the skateboard?

40. A car is travelling at a constant speed along a straight, level road.

    (a) What will happen to the motion of the car if the driver supplies an unbalanced force to it?
    (b) What is the name of the pedal in a car that is used to increase the unbalanced force on the car?

41. A hot air balloon is falling at a steady speed in the air.
    What will happen to the balloon if the balloonist throws a sandbag overboard?

42. What can a free body diagram be used for?
    page 13

43. An athlete of weight 500 N hangs from two rings as shown.
    The angle between each arm and the athlete's body is 30°.

    (a) Draw a free body diagram for the athlete.

    (b) By drawing a vector diagram for the forces acting on the athlete, or otherwise, find the force in each of her arms.

44. A game in a fairground involves a bottle being knocked down from behind by swinging a 1 kg mass on a string at it.

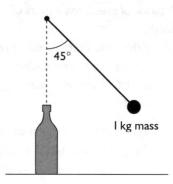

    (a) Draw a free body diagram for the 1 kg mass when it is pulled out by a horizontal force to 45° as shown.

    (b) By drawing a vector diagram for the forces acting, or otherwise, find the horizontal force needed to pull the mass out this far.

45. State what is meant by the **resultant** of a number of forces. *page 13*

46. A tanker is being pulled along in the water by two tugs. Each tug exerts a force of 50 000 N on the tanker. The angle between the two chains joining the tugs to the tanker is 90°. Calculate the resultant force acting on the tanker.

Newton's Second Law of Motion

47. How is the unit of force, the newton, defined?
    page 14

48. (a) State the relationship between acceleration, mass and unbalanced or resultant force, using the usual symbols.
    (b) Give the units for each of the quantities in this expression. *page 14*

49. A trolley of mass 0.75 kg is acted upon by an unbalanced force of 3 N. Calculate its acceleration.

50. Calculate the unbalanced force that will produce an acceleration of $5\,m/s^2$ in a mass of $2\,kg$.

51. An unbalanced force of $1500\,N$ causes a car to decelerate at $2\,m/s^2$.
    Calculate the mass of the car.

52. A parachutist and his parachute have a total mass of $100\,kg$.

    (a) Calculate their combined weight.
    (b) The parachutist has a downward acceleration of $0.1\,m/s^2$.
        Calculate the unbalanced force acting.
    (c) Calculate the air resistance acting on the parachute, and draw a diagram showing all the forces acting.

53. A booster rocket motor is fired on a spacecraft which is far away from any planets. The rocket motor supplies a thrust (a force) of $2000\,N$.
    If the mass of the spacecraft is $5000\,kg$, calculate the acceleration produced in the rocket.

54. (a) Explain why a rocket motor does not need to be kept on all the time while the rocket is moving far away from any planets.
    (b) What would happen to a rocket in space if the rocket motor was fired?

The equivalence of acceleration due to gravity and gravitational field strength

55. Copy and complete the following statements, using words or phrases chosen from the following list:
    acceleration due to gravity;
    gravitational field strength;
    mass;
    weight.

    (a) The quantity of matter in an object is known as its _____ .
    (b) The force of gravity acting on an object is known as its _____ .
    (c) The ratio of weight to mass for an object close to the surface of a planet is known as that planet's _____ .
    (d) Although they have different units, two quantities are equivalent to each other. These quantities are gravitational field strength and _____ . *page 11*

56. Explain how acceleration due to gravity and the gravitational field strength are both equal.
    page 15

Projectiles

57. If an object is projected horizontally, it does not continue to move horizontally.

    (a) Describe the path it takes.
    (b) Explain what causes it to follow this path.
    page 15

58. The path of a projectile can be treated as two independent motions.

    (a) Describe and explain the horizontal motion of a projectile.
    (b) Describe and explain the vertical motion of a projectile. *page 15*

59. There is only one quantity that is common to the vertical and horizontal motions of a projectile. What is it? *page 15*

60. A food parcel is dropped from a helicopter which is flying horizontally at a velocity of $50\,m/s$. If the parcel takes $4\,s$ to reach the ground and air resistance can be ignored, calculate:

    (a) the horizontal velocity of the parcel just as it reaches the ground;
    (b) the horizontal distance travelled by the parcel;
    (c) the vertical velocity of the parcel just as it reaches the ground;
    (d) the height of the helicopter when the parcel was dropped.

61. By considering the motion of a projectile, explain how a satellite remains in orbit.
    page 16

62. Describe and explain Newton's thought experiment. *page 16*

Momentum and energy

Newton's Third Law of Motion

1. State **Newton's Third Law**. *page 17*

2. What causes a rocket in space to move forward? *page 17*

3. Explain the following situations using the rule: 'A exerts a force on B, B exerts an equal but opposite force on A.'
   In each case, draw a diagram with forces on it to help your explanation and say which object you mean as A and which object you mean as B.

(a) Explain how a car moves forward along a road.

(b) Explain how a chair can support a person sitting on it.

(c) Explain how a rocket moves forward in space.

4. What is meant by **'Newton Pairs'**?   *page 17*

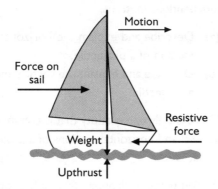

5. Identify the 'Newton Pairs' acting on the boat shown.

6. Although they are different, 'Newton Pairs' are often confused with balanced forces.
Explain clearly the difference between balanced forces and 'Newton Pairs'.   *page 17*

Momentum

7. (a) How is the quantity **momentum** calculated?

(b) What is the unit that is used to measure momentum?   *page 17*

8. Explain the significance of the quantity momentum. Use collisions of cars to illustrate your explanation.   *page 17*

9. Which type of quantity is momentum, a scalar quantity or a vector quantity?   *page 17*

Conservation of momentum

10. State the **Law of Conservation of Momentum** as it relates to two objects moving in one direction. Use a labelled diagram to define all quantities used.   *page 17*

11. When parking, a car of mass 1000 kg travelling at 0.51 m/s bumps into a stationary car of mass 700 kg. They both move off together in the original direction.
Calculate the velocity at which they move off.

12. A hammer of mass 2 kg moving at 10 m/s is used to drive a stake of mass 3 kg into soft ground.

If the hammer remains in contact with the stake, calculate their combined velocity immediately after making contact.

13. In a game of snooker, the cue ball strikes another ball and the second ball moves off in the original direction of the cue ball.
If both balls have the same mass and the cue ball remains stationary after the collision, show that the second ball moves off with the same velocity as the cue ball originally had.

Work and energy

14. (a) What is the symbol that is used for all types of **energy**?

(b) What is the unit, and its abbreviation, that is used to measure energy?   *page 18*

15. Copy and complete:
When work is done by a force, _____ is transferred and the amount of work that is done is a measure of the _____ _____ .
page 18

16. (a) There are two symbols that are often used to denote **work done**. State one of them.

(b) What is the unit, and its abbreviation, that is used to measure work done?   *page 18*

17. (a) State the relationship between work done, force and distance.

(b) Give the units that are used to measure each of the quantities in this expression.
page 18

18. A force of 5 N is used to move a box a distance of 3 m along a bench.
Calculate the amount of work done by the force.

19. Calculate the force that is needed to transfer 1000 J of energy a distance of 20 m.

20. 200 J of work are used up by a force of 40 N on an object.
Calculate the distance that the force moves the object in the direction of the force.

Power

21. (a) What is the symbol that is used for **power**?

(b) What is the unit, and its abbreviation, that is used to measure power?   *page 18*

22. (a) State the relationship between power, work done and time.
    (b) Give the units that are used to measure each of the quantities in this expression.
    page 18

23. A force transfers 360J in 1 minute. Calculate the power involved.

24. A person of mass 60kg runs up a flight of stairs in 10s. The vertical height of the stairs is 5m.

    (a) Calculate the weight of the person.
    (b) Calculate the amount of work done.
    (c) Calculate the power developed by the person in running up the stairs.

25. A small electric motor lifts a mass of 0.3kg at a velocity of 20cm/s.
    Calculate the power developed by the motor.

Potential energy

26. What type of energy is gained by an object when work is done to lift it up? *page 19*

27. Copy and complete:
    The **work done against gravity** is equal to the increase in _____ _____ _____ of an object. *page 19*

28. Under what circumstances is work done on an object *by* gravity? *page 19*

29. (a) What is the symbol that is used for **gravitational potential energy**?
    (b) What is the unit, and its abbreviation, that is used to measure gravitational potential energy? *page 19*

30. (a) State the relationship which is used to calculate gravitational potential energy.
    (b) Give the units that are used to measure each of the quantities in this expression.
    page 19

31. A 50kg bag of cement is raised 1.5m on to the back of a lorry. Calculate the gravitational potential energy gained by the bag of cement.

Kinetic energy

32. What type of energy does an object have because of its motion? *page 19*

33. Copy and complete:
    The greater the _____ and/or the greater the _____ of a moving object, the greater is its kinetic energy. *page 19*

34. (a) What is the symbol that is used for **kinetic energy**?
    (b) What is the unit, and its abbreviation, that is used to measure kinetic energy?
    page 19

35. (a) State the relationship between kinetic energy, mass and speed.
    (b) Give the units that are used to measure each of the quantities in this expression.
    page 19

36. Calculate the amount of kinetic energy a trolley of mass 0.75kg has when it is travelling at 2m/s.

37. Show by calculation which has the greater amount of kinetic energy – a car of mass 800kg travelling at 26m/s (about 60 miles per hour) or a lorry of mass 3 tonnes travelling at 13m/s (about 30 miles per hour).
    (One tonne = 1000kg)

Machines and efficiency

38. Describe what is meant by an **energy transformation**. *page 20*

39. Give the equation that is used to calculate the efficiency of an energy transformation.
    page 21

40. Calculate the percentage efficiency of an electric kettle which has an input power rating of 3000W and an output power of 2700W.

41. A small electric motor uses 10J of energy in lifting a mass of 100g a distance of 2m. Calculate the percentage efficiency of the electric motor.

Heat
Heat and temperature

1. What is meant by the **temperature** of an object? *page 20*

2. What is the name of the scale used to measure temperature? *page 20*

3. Copy and complete:

    _____ is a measure of how hot or cold an object is. It is measured in _____ .
    Heat is a form of _____ and is measured in
    _____ .

    Putting _____ into an object usually makes its _____ increase. *page 20*

4. Copy and complete, using the words 'higher' and 'lower' in the spaces:

    Heat travels from a region of _____ temperature to a region of _____ temperature. *page 20*

5. Does 1 kg of copper need the same amount of energy as 1 kg of aluminium to raise its temperature by 1°C? *page 20*

6. Compare the amount of energy needed to raise the temperature of 2 kg of copper by 1°C with that needed to do the same to 1 kg of copper. *page 20*

7. Compare the amount of energy needed to raise the temperature of 1 kg of copper by 1°C with that needed to raise the temperature by 2°C. *page 20*

8. Compare the amount of energy needed to raise the temperature of 1 kg of copper from 9°C to 10°C with that needed to raise the temperature from 99°C to 100°C. *page 20*

Specific heat capacity

9. Explain what is meant by the **specific heat capacity** of a substance. *page 21*

10. (a) What is the symbol that is used for specific heat capacity?
    (b) What is the unit, and its abbreviation, that is used to measure specific heat capacity? *page 21*

11. (a) State the equation that links heat to the mass, specific heat capacity and temperature change in a substance.
    (b) Give the units that are used to measure each of the quantities in this equation. *page 21*

12. Calculate the amount of heat needed to increase the temperature of 1 kg of aluminium by 10°C.
    (The specific heat capacity of aluminium is 902 J/kg°C.)

13. Calculate the amount of heat needed to raise the temperature of 2 kg of water from 20°C to 90°C.
    (The specific heat capacity of water is 4180 J/kg°C.)

14. Calculate the increase in temperature of the 5 kg of coolant in the cooling system of a car after it has absorbed 720 kJ of energy from the car engine.
    (The specific heat capacity of the coolant used is 2400 J/kg°C.)

Change of state

15. What are the three **states of matter**? *page 21*

16. What is meant by a **change of state**? *page 21*

17. (a) What two things *can* happen to a body when heat is transferred to it?
    (b) Explain how to decide which of the two things *will* happen in any particular case. *page 21*

18. Copy and complete the diagram, giving the names of the processes missing from the boxes labelling the arrows.

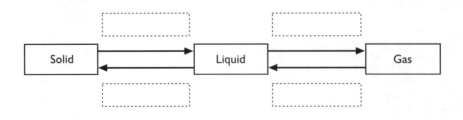

19. What does *not* change when the state of a substance is changed? *page 21*

20. (a) What must be added to a substance to make its state change from a solid to a liquid or from a liquid to a gas?
    (b) What is lost from a substance when its state changes from a gas to a liquid or from a liquid to a solid? *page 21*

Applications of change of state

21. Give *two* examples of applications which involve a change of state. *page 22*

Latent heat

22. (a) What does the word 'latent' mean?
    (b) Hence explain how the term 'latent heat' comes about. *page 22*

23. (a) Give another word which means the same as 'fusion'.
    (b) Give another term which is often used for 'vaporisation'.

24. (a) Explain what is meant by the **latent heat of fusion** of a substance.
    (b) Explain what is meant by the **latent heat of vaporisation** of a substance. *page 22*

25. (a) What is the symbol that is used for **specific latent heat**?
    (b) What is the unit, and its abbreviation, that is used to measure specific latent heat? *page 23*

26. (a) State the equation that links heat to the mass and the specific latent heat of a substance.
    (b) Give the units that are used to measure each of the quantities in this equation. *page 23*

27. Calculate the amount of heat needed to convert 0.5 kg of water at its boiling point into steam. (The specific latent heat of vaporisation of water is 2.26 x 10^6 J/kg.)

28. Calculate the amount of heat given out when 0.5 kg of water freezes to ice at 0°C. (The specific latent heat of fusion of water is 3.34 x 10^5 J/kg.)

29. A lump of ice of mass 1 kg is removed from a freezer at a temperature of −18°C. Calculate the amount of heat which it must be given in order to completely convert it to steam at 100°C.

    (Note that although this problem may be considered difficult, it will be found to be easier if treated logically.)
    (The specific heat capacity of ice is 2100 J/kg°C; the specific heat capacity of water is 4180 J/kg°C; the specific latent heat of fusion of water is 3.34 x 10^5 J/kg; the specific latent heat of vaporisation of water is 2.26 x 10^6 J/kg.)

Conservation of energy

30. State and explain the **principle of conservation of energy**. *page 23*

31. An electric kettle is rated at 2.2 kW and has a capacity of 1.5 litres of water. Calculate how long it would take to increase the temperature of the water from room temperature (20°C) to boiling point, without boiling any water off.
    (The specific heat capacity of water is 4180 J/kg °C; mass of 1 litre of water is 1 kg.)

32. A car of mass 1000 kg, travelling at 4 m/s, is brought to rest by applying the brakes. Calculate the increase in temperature of the brake linings and pads if all of the original kinetic energy of the car is transferred to heat in the brakes.
    (Total mass of brake lining material = 0.5 kg; specific heat capacity of brake lining material = 500 J/kg°C.)

33. A meteorite consists of a lump of iron of mass 3 kg. It enters the Earth's atmosphere at 2000 m/s. Assuming 10% of the kinetic energy of the meteorite is converted into heat (which in fact does not happen), calculate its rise in temperature. (The specific heat capacity of iron is 440 J/kg°C.)

34. Show that, if air resistance can be ignored, the speed of a falling object is independent of its mass and depends only on the height through which it moves.

35. A stone falls from a cliff which is 80 m high.

    (a) If air resistance can be ignored, calculate the speed at which it enters the water at the bottom of the cliff.

    (b) If air resistance *cannot* be ignored, what effect will this have on the speed of the stone as it enters the water?

    (c) In practice, not all of the initial gravitational potential energy is transformed into kinetic energy. Other than kinetic energy, what is the main form of energy produced?

36. (a) What type of energy does a spacecraft have because of its movement?

    (b) What is this energy changed into when the spacecraft re-enters the Earth's atmosphere from space?

    (c) What causes this energy transformation to take place?

37. The orbiter part of the Space Shuttle, the part that returns to Earth after the space mission, has a mass of 70 000 kg. While in orbit, its velocity is about 8000 m/s and at touchdown its velocity is about 100 m/s.

    (a) Calculate the kinetic energy of the orbiter while it is in orbit.

    (b) Calculate the kinetic energy of the orbiter just as it touches down.

    (c) What has happened to the 'lost' kinetic energy between being in orbit and at touchdown?

    (d) The average force needed to stop the orbiter as it travels along the runway between touchdown and coming to rest is 175 kN.
    Calculate the length of the runway needed.

ELECTRICITY AND ELECTRONICS

Circuits

Charge and current

One of the fundamental properties of materials is that of **charge**.

Electrostatics is the study of charges at rest. Simple electrostatics experiments help us to establish the following.

(i) There are two types of charge: positive and negative. The names have no significance other than as labels for the two types of charge.

(ii) All materials are made up of atoms which normally contain equal numbers of positively and negatively charged particles. (If you are unfamiliar with the structure of the atom, see page 77.) When there is no excess charge of either type, the material is said to be uncharged. Some types of materials can be charged, either positively or negatively, by rubbing (causing friction). Since electrons carry negative charges, if electrons are rubbed on to a material it becomes negatively charged, while if a material loses electrons it becomes positively charged. Electrons in a conductor are free to move and so conductors cannot easily be charged since the charge usually flows away as soon as it is placed on the conductor.

(iii) Like or similar charges exert forces which make them repel each other. Unlike charges exert forces which make them attract each other.

When charges move around a circuit, they form an **electric current**. An electric current is a flow of charges, usually negative charges carried by electrons.
Current is the rate of flow of charges.

The symbol for electric charge is Q. The unit that is used to measure electric charge is the coulomb (C).

The symbol for electric current is I. The unit that is used to measure electric current is the ampere (A).

The relationship between charge, current and time is:

charge = current x time
$$Q = I\,t$$
where Q is charge in C
      I is current in A
      t is time in s.

EXAMPLE

Calculate the charge transferred in a circuit which has a current of 0.25 A for 1 hour.

SOLUTION

current = 0.25 A
time = 1 hour = 60 × 60 s = 3600 s
charge = ?
     $Q = I\,t = 0.25 \times 3600 = 900$
charge transferred = 900 C

An **electrical conductor** contains electrons which are free to move and so allows an electric current, which is a flow of charges.

An **insulator** does not allow an electric current because the charges in it are not free to move.

Metals such as gold, silver, copper and aluminium are good conductors.

Materials which are insulators are usually non-metals, examples being plastics such as polythene, pvc and bakelite as well as wood, paper and ceramic materials.

Circuit symbols

When electric circuits are drawn, it is usual to use a set of internationally known and accepted symbols for the components.
Some of these symbols are as follows.

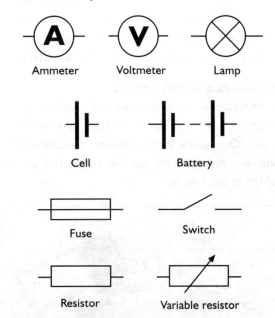

| Ammeter | Voltmeter | Lamp |

| Cell | Battery |

| Fuse | Switch |

| Resistor | Variable resistor |

Voltage and potential difference

The **voltage** of an electrical energy supply (cell or battery) is a measure of the energy given to the charges in a circuit. If a lamp is included in a circuit where the voltage across it can be altered, then the brightness of the lamp will increase as the voltage increases. This is because the brightness of the lamp increases as the charges flowing through it are given more energy.

The symbol for voltage is V. The unit that is used to measure voltage is the volt (V).

As the charges in a circuit move through each component in the circuit they lose their energy. Because of this there is a difference in energy or potential across each component. The voltage across a component is known as **potential difference**.

Note the similarities and the differences between the two terms 'potential difference' and 'voltage'. Both relate to the energy carried by the charges in a circuit. 'Voltage' is the term used to describe the energy *given* to the charges by the battery, cell or other source. 'Potential difference' refers to the difference in energy levels of the charges across a component and so describes the energy *changed to other forms* as the charges pass through the component.

Voltage and current are not the same but are often confused. Voltage is a measure of the amount of energy given to the charges in a circuit. Current is a measure of the rate at which charges flow round a circuit.
The voltage applied to a circuit causes the charges in the circuit to flow and this flow of charges is the current.

Resistance
Resistance is a measure of the opposition to the flow of charges in a circuit. The symbol for resistance is R. The unit that is used to measure resistance is the ohm (Ω). A component in a circuit which has the property of resistance is called a **resistor**.

Ammeters and voltmeters
An ammeter is used to measure current. It is connected into a circuit by making a gap in the circuit. One extra lead is needed to connect an ammeter. An ammeter can be connected at any position in the circuit.

An ammeter and voltmeter

A **voltmeter** is used to measure voltage or potential difference. It is connected across a component in the circuit using two extra leads.

In the circuit shown in the diagram the ammeter measures the current through the lamp and the voltmeter measures the potential difference across the lamp.

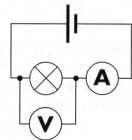

When the resistance in a circuit is increased, the current in the circuit decreases. This can be verified by using the following circuit.

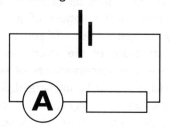

By inserting resistors of known values in this circuit, you can see that the current decreases as the value of the resistor increases, even when the supply voltage is kept constant.

Series circuits
A **series circuit** is one where there is only one path for the charges to follow. All of the components are connected one after the other.

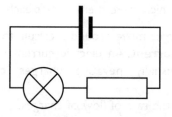

A series circuit

The current is the same at all points in a series circuit. This can be shown using the following circuit.

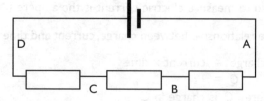

Make a gap at positions A, B, C and D in turn and insert an ammeter in the gap. The meter shows the same reading of current at all positions.

The sum of the potential differences across the components in a series circuit is equal to the voltage of the supply.

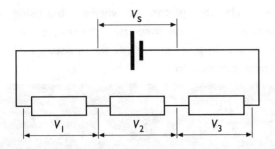

In the circuit shown, V_1, V_2 and V_3 are the potential differences across the three resistors and V_s is the voltage of the supply. When these four voltages are measured you should find that

$$V_1 + V_2 + V_3 = V_s$$

Parallel circuits

A **parallel circuit** is one which has more than one branch or path for charges to follow. There will always be at least two junctions of wires in a parallel circuit.

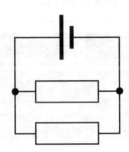

A parallel circuit

The current drawn from the supply in a parallel circuit is equal to the sum of the currents in the individual branches.

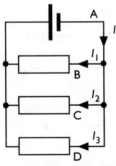

In the circuit shown, place ammeters in positions A, B, C and D to measure the supply current I and the three currents I_1, I_2 and I_3 through the three resistors. You should find that

$$I = I_1 + I_2 + I_3$$

The potential difference across each component in a parallel circuit is the same for each component. If the components are connected directly across the supply, the voltage is also equal to the supply voltage. In the circuit shown, voltmeters used to measure the voltage of the supply, V, and the potential differences

V_1, V_2 and V_3 across the individual resistors would give four readings which are the same.

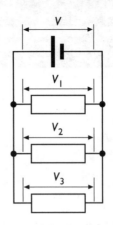

Resistance and Ohm's Law

The ratio V/I for a resistor remains approximately constant when the current in the resistor changes. This ratio for a resistor is called **resistance**. Resistance is a property of the resistor, not the current through it nor the potential difference across it.

The relationship between current, voltage (potential difference) and resistance for a resistor is:

$$\text{resistance} = \frac{\text{voltage}}{\text{current}} \qquad R = \frac{V}{I}$$

This relationship can be verified as follows. Measure the resistance R of a resistor using an ohmmeter. Put the resistor into the circuit as shown.

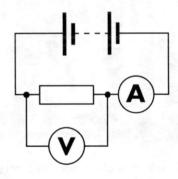

Read the values of current and potential difference from the ammeter and the voltmeter as the number of cells in the battery is altered.

You should find that the ratio V/I in each case is approximately constant and is equal to the resistance of the resistor.

The relationship between current, potential difference and resistance is formally known as **Ohm's Law** for a conductor:

The current in a conductor at constant temperature is directly proportional to the potential difference across it.

EXAMPLE

A car headlamp bulb takes a current of 3 A from the 12V car battery.
Calculate the resistance of the bulb.

SOLUTION

current = 3 A
voltage = 12 V
resistance = ?

$$\text{resistance} = \frac{\text{voltage}}{\text{current}} \qquad R = \frac{V}{I} = \frac{12}{3} = 4$$

resistance of bulb = 4 Ω

Resistors in series and parallel

When resistors are connected in series as shown,

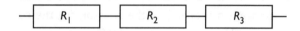

the total effective resistance, R_T, is given by

$$R_T = R_1 + R_2 + R_3$$

This relationship can be verified by using an ohmmeter to measure the resistance of the individual resistors as well as the resistance of the series combination.

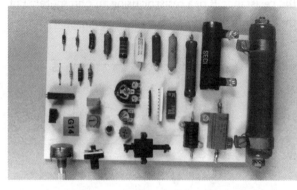

A selection of electrical resistors

When resistors are connected in parallel as shown,

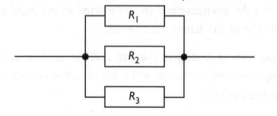

the total effective resistance R_T is given by

$$\frac{1}{R_T} = \frac{1}{R_1} + \frac{1}{R_2} + \frac{1}{R_3}$$

This relationship can be verified by using an ohmmeter to measure the resistance of the individual resistors as well as the resistance of the parallel combination.

EXAMPLE

Calculate the total resistance in the following circuit.

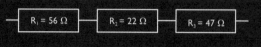

SOLUTION

For resistors in series:
$$R_T = R_1 + R_2 + R_3$$
$$R_T = 56 + 22 + 47$$
$$R_T = 125 \,\Omega$$
total resistance = 125 Ω

EXAMPLE

Calculate the total resistance in the following circuit.

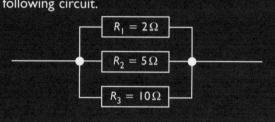

SOLUTION

For resistors in parallel:

$$\frac{1}{R_T} = \frac{1}{R_1} + \frac{1}{R_2} + \frac{1}{R_3}$$

$$\frac{1}{R_T} = \frac{1}{2} + \frac{1}{5} + \frac{1}{10}$$

$$\frac{1}{R_T} = \frac{5+2+1}{10} = \frac{8}{10}$$

$$R_T = \frac{10}{8} = 1.25$$

total resistance = 1.25 Ω

The potential divider

A **potential divider** circuit consists of a number of resistors connected across a supply.

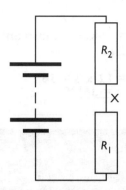

Potential divider circuit

The potential (voltage) at point X is a fixed fraction of the supply voltage and is determined by the values of the two resistors, R_1 and R_2.

The following relationships hold for the potential divider circuit shown.

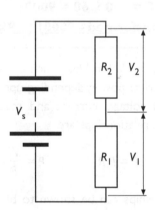

$$V_s = V_1 + V_2$$

$$V_1 = V_s \times \frac{R_1}{R_1 + R_2}$$

$$\frac{V_1}{V_2} = \frac{R_1}{R_2}$$

$$V_2 = V_s \times \frac{R_2}{R_1 + R_2}$$

EXAMPLE

A potential divider consists of a resistor R_1 of value $1000\,\Omega$ in series with a resistor R_2 of value $1500\,\Omega$. It is connected across a supply voltage of 2.5V.
Draw the circuit diagram and calculate the potential difference across R_1.

SOLUTION

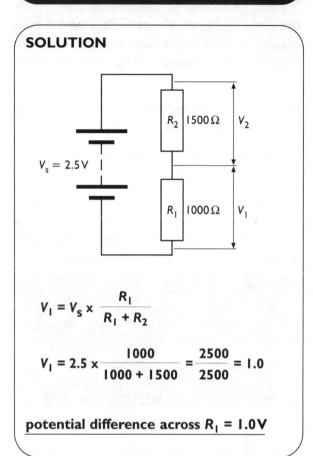

$$V_1 = V_s \times \frac{R_1}{R_1 + R_2}$$

$$V_1 = 2.5 \times \frac{1000}{1000 + 1500} = \frac{2500}{2500} = 1.0$$

potential difference across R_1 = 1.0V

If one of the resistors is a variable resistor, then the potential at the junction can be varied. This is used to adjust the sensitivity of transistor circuits (see page 49).

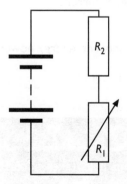

Variable potential divider circuit

Electrical energy

Energy and power in electrical circuits

When there is a current in a component in a circuit the electrical **energy** is transformed or changed into other forms, often heat and light.

The electrical energy from the power supply is used up in moving the charges round the circuit.

The useful energy transformation in a lamp is from electrical to light. This means that, as the current through a lamp increases, the brightness of the lamp also increases.

The symbol for energy is *E*. The unit that is used to measure energy is the joule (J).

Power is the rate at which energy is converted from one form into another.

The **power rating** of an electrical component is the rate at which it transforms electrical energy. The power rating of an electrical appliance is sometimes referred to as the 'wattage' of the appliance.

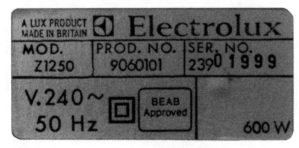

A rating plate

The symbol for power is *P*. The unit that is used to measure power is the watt (W). One watt is one joule per second.

The relationship between power, current and voltage is:

power = voltage x current
$$P = VI$$

The relationship between energy, power and time is:

energy = power x time
$$E = Pt$$

EXAMPLE

A car headlamp bulb takes a current of 3A from the 12V car battery.
Calculate the power rating of the bulb.

SOLUTION

current = 3A
voltage = 12V
power = ?
    power = voltage x current
        $$P = VI$$
        $$P = 12 \times 3 = 36$$
<u>**power rating = 36W**</u>

EXAMPLE

Calculate the amount of energy transformed in a 150W light bulb every minute.

SOLUTION

power = 150W
time = 1 minute = 60s
energy = ?
    energy = power x time
        $$E = Pt$$
        $$E = 150 \times 60 = 9000$$
<u>**energy transformed = 9000J (= 9kJ)**</u>

There are other relationships which can be used to calculate electrical power, depending upon which of the quantities voltage, current and resistance are known. These relationships are:

$$P = VI \qquad P = I^2 R \qquad P = \frac{V^2}{R}$$

These relationships can be shown to be equivalent to each other as follows.

$P = V \times I$ and $V = I \times R$
so $P = (I \times R) \times I$
so $P = I^2 R$

$P = V \times I$ and $I = \dfrac{V}{R}$
so $P = V \times \dfrac{V}{R}$

so $P = \dfrac{V^2}{R}$

EXAMPLE

A 60W light bulb takes a current of 0.25A. Calculate its resistance.

SOLUTION

power = 60 W
current = 0.25 A
resistance = ?

$P = I^2 R$

so $60 = 0.25^2 \times R$

so $R = \dfrac{60}{0.25^2} = 960$

resistance of light bulb = 960 Ω

EXAMPLE

A bulb used in a car with a 12 V electrical system has a resistance of 18 Ω. Calculate the power of the bulb.

SOLUTION

voltage = 12 V
resistance = 18 Ω
power = ?

$P = \dfrac{V^2}{R} = \dfrac{12 \times 12}{18} = 8$

power of bulb = 8 W

Fuses and appliances

The two sizes of fuse recommended for use in mains plugs are 3 A and 13 A. The power rating of an appliance can be used to choose the correct fuse to use in a mains plug:

- for appliances up to 700 W, use a 3 A fuse;
- for appliances greater than 700 W, use a 13 A fuse.

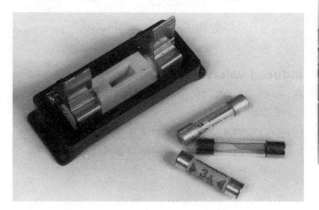

Alternatively, the fuse needed in the plug of an appliance can be determined by calculating the current taken by the appliance:

- if the current is less than 3 A, use a 3 A fuse;
- if the current is 3 A or greater, use a 13 A fuse.

EXAMPLE

Calculate the current taken by a 920 W hair dryer when it is connected to the 230V mains supply. Use your answer to determine the correct fuse to use in the mains plug ofthe hair dryer.

SOLUTION

power = 920 W
voltage = 230 V
current = ?

power = current x voltage

$I = \dfrac{P}{V} = \dfrac{920}{230} = 4$

current taken = 4 A

This current is greater than 3 A, so use a 13 A fuse.

Electrical appliances change electrical energy from the mains supply into another more useful form of energy, usually heat, light, kinetic or sound.

In a lamp, the electrical energy is transformed into heat and light.

In an electric heater, the energy transformation takes place in the element. The element is made from resistance wire. Appliances used in the home in which electrical energy is transformed into heat include: electric fire, toaster, kettle, electric cooker, soldering iron.

Electrical appliances

The heating element of an electric fire has a far higher resistance than the flex which connects the fire to the mains. So the rate at which electrical energy is transformed to heat ($I^2 R$) in the heating element is far greater than in the flex even although the current is the same in both.

Direct current and alternating current

The term **direct current** (d.c.) describes the type of current where the charges flow the same way all the time.

The term **alternating current** (a.c.) describes the type of current where the charges go backwards and forwards many times a second.

The **mains supply** or **battery** is the source of electrical energy in an electrical circuit.

- The mains supply is a.c.
- The output from a battery is d.c.

The traces seen on an oscilloscope connected to a battery (d.c.) and a low voltage a.c. supply are as shown below.

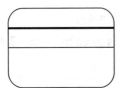

d.c. from a battery

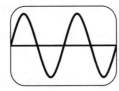

a.c. from the mains supply

The **frequency** of the mains supply in Britain is 50 Hz (cycles per second).

The **quoted value** of an alternating voltage (the voltage of an a.c. supply) is less than the **peak value** of the same voltage. This is because the quoted value of an alternating voltage is the r.m.s. (root mean square) value (the effective average) of the mains supply so it is smaller than the peak value of the same voltage.

A d.c. supply and an a.c. supply of the same quoted value will supply the same power to a given resistor.

The **declared voltage** or the **quoted value** of the mains supply in Britain is 230 V.

Electromagnetism

Magnets and magnetic fields

A **magnet** has the property of attracting some metals, in particular iron and nickel. Such metals are known as **magnetic materials**. Magnets can either occur naturally, for example lodestone, or they can be made from magnetic materials. A freely suspended magnet forms the basis of a compass since it will always come to rest pointing in a north–south direction. This is because the Earth acts like a gigantic bar magnet and the ends (poles) of magnets either attract or repel each other.

A **magnetic field** is the region around a magnet where the magnet exerts a force. A magnetic field can be represented by field lines with arrowheads showing the way a small compass needle would line up in the field.

An electric current in a wire produces a magnetic field in the space around the wire.

A circular magnetic field exists around a straight, current-carrying wire.

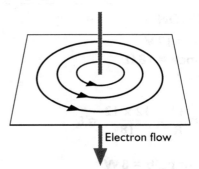

Electron flow

Magnetic field around a wire

The shape of the magnetic field around a current-carrying coil is as shown.

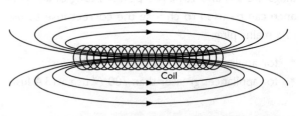

Coil

Magnetic field around a coil

Induced voltage

A voltage will be **induced** or generated in a conductor when:

(i) the conductor moves in a magnetic field;

(ii) the conductor is in a changing magnetic field.

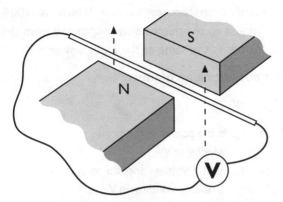

Moving a conductor in a magnetic field

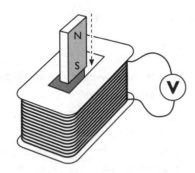

Moving a magnet into or out of a coil of wire (changing magnetic field)

There are three factors which affect the size of the voltage induced in a conductor.

- Increasing the strength of the magnetic field increases the size of the induced voltage.

- Increasing the number of turns on the coil increases the size of the induced voltage.

- Increasing the relative speed of the magnet and the coil increases the size of the induced voltage.

Transformers

Transformers are used to change the size or magnitude of an alternating voltage.

A transformer consists of two coils of wire, electrically unconnected, both wound on to an iron core. One coil, called the **primary winding**, has an alternating voltage applied to it; the other coil, the **secondary winding**, supplies an alternating voltage to a circuit.

An alternating voltage applied to the primary winding of a transformer causes a changing magnetic field in the iron core. This changing magnetic field induces an alternating voltage in the secondary winding of the transformer. Since a

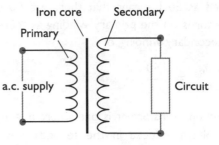

Transformer symbol

changing magnetic field is needed for a voltage to be induced in a conductor, transformers only work with alternating voltages.

A transformer

The input and output voltages of a transformer are related through the transformer equation:

$$\frac{V_p}{V_s} = \frac{n_p}{n_s}$$

where V_p is the voltage applied to the primary winding in volts

V_s is the voltage induced in the secondary winding in volts

n_p is the number of turns on the primary winding

n_s is the number of turns on the secondary winding.

$\frac{n_p}{n_s}$ is sometimes known as the turns ratio of the transformer. So, for example, a turns ratio quoted as 50:1 means that there are 50 times as many turns on the primary winding as there are on the secondary winding, or

$$\frac{n_p}{n_s} = \frac{50}{1}$$

A 'step-up' transformer is one where $n_s > n_p$ and so the voltage induced in the secondary winding is greater than the voltage applied to the primary winding.

A 'step-down' transformer is one where $n_s < n_p$ and so the voltage induced in the secondary winding is less than the voltage applied to the primary winding.

EXAMPLE

A transformer has an input voltage of 11000 V. There are 55000 turns on its primary winding and 1150 turns on its secondary winding.

(a) Is the transformer a step-up or a step-down transformer?
(b) Calculate the output voltage of the transformer.

SOLUTION

$V_p = 11000 V$

$n_p = 55000$

$n_s = 1150$

(a) **Since $n_s < n_p$, this is a step-down transformer.**

(b) **Output voltage:**

$$\frac{V_p}{V_s} = \frac{n_p}{n_s}$$

so $V_s = \dfrac{n_s}{n_p} \times V_p$

$$= \frac{1150}{55\,000} \times 11\,000 = 230$$

output voltage = 230V

An **ideal transformer** is one which is 100% efficient and so the output power obtained from the transformer is the same as the input power to it.

Input power to a transformer:

$$P_{in} = V_p I_p$$

where P_{in} is the power in the primary winding in W

V_p is the voltage applied to the primary winding in V

I_p is the current in the primary winding in A.

Output power from a transformer:

$$P_{out} = V_s I_s$$

where P_{out} is the power in the secondary winding in W

V_s is the voltage generated in the secondary winding in V

I_s is the current in the secondary winding in A.

The following relationship only holds for an ideal transformer.

$$\text{output power} = \text{input power}$$
$$P_{out} = P_{in}$$
$$V_s I_s = V_p I_p$$

$$\frac{V_p}{V_s} = \frac{I_s}{I_p}$$

EXAMPLE

Calculate the current in the primary winding of an ideal mains transformer when a 24V electric drill connected to the secondary draws a current of 2.3A.

SOLUTION

$V_s = 24 V$

$I_s = 2.3 A$

$V_p = 230 V$ **(mains transformer)**

$$\frac{V_p}{V_s} = \frac{I_s}{I_p}$$

so $I_p = I_s \times \dfrac{V_s}{V_p} = 2.3 \times \dfrac{24}{230} = 0.24$

current in primary = 0.24A

Generating station and transmission lines

Transmission lines and the National Grid

The electricity produced by power stations is generated at 25 kV.

This is increased to 132 kV, 275 kV or 400 kV by step-up transformers, and fed into the **National Grid** system to be transmitted over long distances. (The reason for stepping up the voltage is explained later.)

Step-down transformers in sub-stations supply users with electricity at voltages suitable for their needs as follows:

- heavy industry needs a 33 kV supply;
- rail electrification needs a 25 kV supply;
- light industry and hospitals need an 11 kV supply;
- homes, shops and offices need a 230 V supply.

The transmission of electricity from the power station where it is generated to the consumer where it is used can be modelled as shown below.

When electricity is transmitted over long distances, power is lost because of the resistance of the transmission lines.

The power loss in the transmission lines can be calculated from:

$$P = I^2 R$$

where P is power (loss) in W

I is transmission line current in A

R is resistance of transmission lines in Ω.

The power loss in transmission lines is kept as low as possible by the following means:

- transmitting a given amount of power at a higher voltage and so at a lower current – this reduces the value of I in the equation $P = I^2 R$.
- using wire with a low resistance for the transmission lines – this reduces the value of R in the equation $P = I^2 R$.

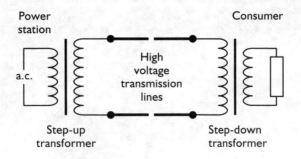

Model power transmission system

Electronic components

Input, process and output

An electronic system can be split into **input**, **process** and **output**.

Output devices

An **output device** in an electronic system transforms the electrical energy from the process part of the system into a different form of energy. Some output devices, the energy transformations involved in them and examples of applications of them are given in the table below.

A **seven–segment display** is an output device consisting of seven LEDs, each shaped like a bar, arranged as a figure of eight. Different numbers can be produced on a seven–segment display by switching on different combinations of LEDs.

A **relay** is an electromagnetic switch. When the coil is activated, the switch in the relay is turned on or off.

A **solenoid** is an output device that moves a metal bar in a straight line using electromagnetism.

Output device	Energy transformation	Example of application
light-emitting diode (LED)	electrical into light	as a 'power-on' indicator for a personal stereo
lamp	electrical into light	as a visible light warning signal on a burglar alarm
seven-segment display	electrical into light	as a television channel indicator
loudspeaker	electrical into sound	in a stereo hi-fi system
relay	electrical into kinetic	to switch on a high current motor with a small current
solenoid	electrical into kinetic	to push a packet of sweets to the hatch in a vending machine
buzzer / bell	electrical into sound	as an audible warning device in an incubator
electric motor	electrical into kinetic	to turn a conveyor belt round in a factory

Electronic devices

The LED

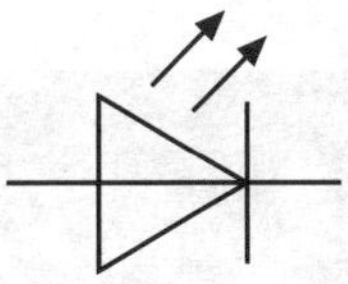

LED symbol

An LED lights only when it is connected the correct way round in a circuit. The LED symbol is rather like an arrowhead. When the arrowhead points towards the negative terminal of the supply, the LED is connected the correct way round. When the connections to the LED are reversed, the LED will not emit light because it will not allow a current through.

This diagram shows a circuit which allows an LED to light.

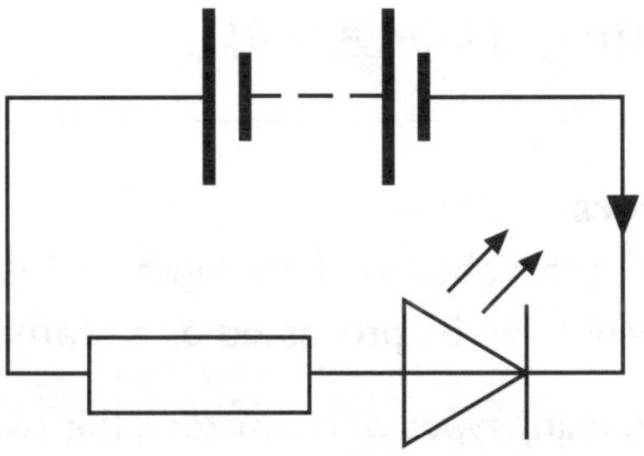

A resistor is needed in series with the LED to limit the current, so that the LED is not damaged. The resistance of the resistor can be calculated by applying the formula $V = IR$, using the appropriate values of V, I and R.

EXAMPLE

An LED is used in a circuit with a 12 V supply. The potential difference across the LED is 2 V and the current through it is 10 mA. Calculate the value of series resistor that is needed.

SOLUTION

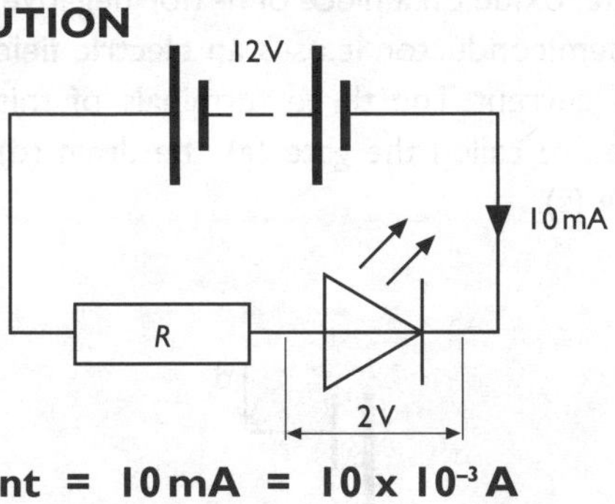

current = 10 mA = 10×10^{-3} A
potential difference across resistor = (12 – 2) V = 10 V

$$R = \frac{V}{I} = \frac{10}{10 \times 10^{-3}} = 1000$$

<u>resistance of series resistor = 1000 Ω</u>

Both the filament lamp and the LED transform electrical energy into light. However, there are differences between them:

- the filament lamp uses more energy than the LED in a given time;
- because of this, the filament lamp gives out more light than the LED;
- a filament lamp gives out white light while an LED gives out light of one colour only;
- a filament lamp also gives out heat while an LED does not.

Input devices

An **input device** in an electronic system converts some form of energy into electrical energy for processing in the next stage of the electronic system. Some input devices, the energy transformations involved in them and examples of applications of them are given in the table below.

Input device	Energy transformation	Example of application
microphone	sound into electrical	in a heartbeat monitor
thermocouple	heat into electrical (produces a small voltage when heated)	in the measurement of high temperatures, for example a gas fire flame sensor
solar cell	light into electrical (produces a voltage when light falls on it)	as power supplies for spacecraft and satellites; also used in light meters

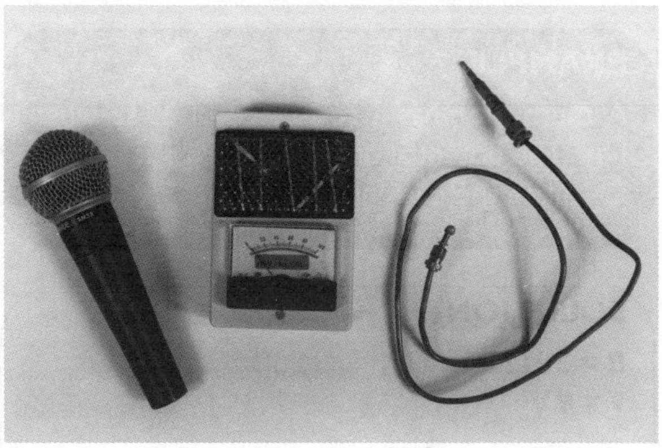

Input devices (microphone, thermocouple, solar cell)

UNIT 2

The thermistor and the LDR

A **thermistor** (thermal resistor) has a resistance which changes as its temperature changes. Normally the higher the temperature, the lower its resistance. Thermistors are used in the measurement of temperatures.

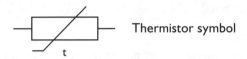

Thermistor symbol

A **Light Dependent Resistor** (LDR) has a resistance which changes as the level of light falling on it changes. Normally, the greater the light intensity, the lower the LDR's resistance. LDRs can be used as automatic light controls and light sensors in alarms, such as burglar alarms.

LDR symbol

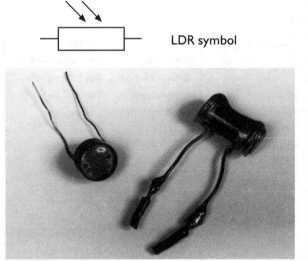

Thermistor and LDR

EXAMPLE

A thermistor has a resistance of 30 Ω when hot. The potential difference across it is 9 V. Calculate the current in it.

SOLUTION

$R = 30\,\Omega$
$V = 9\,V$

$$I = \frac{V}{R} = \frac{9}{30} = 0.3$$

current in thermistor = 0.3 A

EXAMPLE

An LDR has a current of 10 mA in it and a potential difference of 4 V across it. Calculate its resistance.

SOLUTION

$I = 10\,mA = 10 \times 10^{-3}\,A$
$V = 4\,V$

$$R = \frac{V}{I} = \frac{4}{10 \times 10^{-3}} = 400$$

resistance of LDR = 400 Ω

Transistors

Electrical signals obtained from input devices in electronic circuits can be processed by a **transistor**.

There are many types of transistors, the two largest groups being bipolar junction transistors and field effect transistors. The names come from how they are constructed or how they operate and are largely unimportant.

There are different types of transistors within each of these groups. For this Course, it is necessary to draw and identify the circuit symbols for two only: the n-channel enhancement MOSFET and the NPN transistor.

n-channel enhancement MOSFET. The name of this device tells an engineer how it is constructed and how it operates. The word MOSFET comes from the first letters of Metal Oxide Semiconductor Field Effect Transistor. This type of transistor has a layer of metal oxide on a piece of n- (for negative charges) type semiconductor. It uses an electric field to control a current. The three terminals of this type of device are called the gate (g), the drain (d) and the source (s).

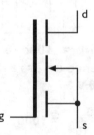

n-channel enhancement MOSFET symbol

NPN transistor. This type of junction transistor consists of a 'sandwich' of p-type semiconductor between two areas of n-type. Its three terminals are called the base (b), the collector (c) and the emitter (e).

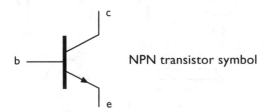

NPN transistor symbol

The transistor as a switch

A transistor can be used as a switch. An NPN transistor can be switched on when the voltage at the base is high enough (about 0.7 V). When it is switched on, it conducts and allows a current between the collector and the emitter.

EXAMPLE

State the purpose of the simple transistor switching circuit shown and explain how the circuit operates.

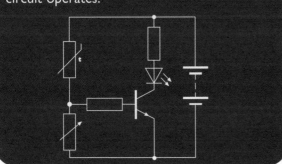

SOLUTION

This circuit lights the LED when the temperature of the thermistor rises above a certain value.

When the temperature of the thermistor rises, its resistance decreases. This causes the potential at the base of the transistor to rise. When this potential reaches about 0.7 V, the transistor conducts and the LED emits light.

The variable resistor controls the point at which the potential at the base is sufficient to switch on the transistor. This adjusts the sensitivity of the circuit to temperature.

Amplifiers

An **amplifier** increases the amplitude or strength of an electrical signal. Amplifiers play an important part in many electrical devices, such as:

> radio; intercom; hi-fi system; personal stereo; baby alarm; public announcement (PA) system; television.

The frequency of the output signal from an audio amplifier is the same as that of the input signal to the amplifier.

The amplitude of the output signal (or the output voltage) from an audio amplifier is greater than that of the input signal (or the input voltage) to the amplifier.

The **voltage gain** of an amplifier is the number of times the output voltage of the amplifier is greater than the input voltage to the amplifier.

$$\text{voltage gain} = \frac{\text{output voltage}}{\text{input voltage}}$$

(Input and output voltages are both measured in volts. Voltage gain, being a ratio, has **no** units.)

EXAMPLE

Calculate the voltage gain of an amplifier which has an input voltage of 2 mV and an output voltage of 0.4 V.

SOLUTION

input voltage $= 2\,\text{mV} = 2 \times 10^{-3}\,\text{V}$
output voltage $= 0.4\,\text{V}$

$$\text{voltage gain} = \frac{\text{output voltage}}{\text{input voltage}}$$

$$= \frac{0.4}{2 \times 10^{-3}} = 200$$

voltage gain = 200

To measure the voltage gain of an amplifier, the input signal to the amplifier is fed to an oscilloscope and its amplitude calculated by measuring the height of the trace and using the voltage sensitivity setting of the oscilloscope. In a similar way the amplitude of the output voltage is also calculated using the oscilloscope. The voltage gain is then found by using the voltage gain equation above.

ELECTRICITY AND ELECTRONICS QUESTIONS

Circuits

Charge and current

1. (a) How many types of **charge** are there?
    (b) Name them.   *page 35*

2. Explain why most objects are normally uncharged, and use this to explain how an object can become charged.   *page 35*

3. Make a statement about the attraction and repulsion of charges.   *page 35*

4. Which kind of charge is carried on an electron?   *page 35*

5. (a) What is the symbol for electric charge?
    (b) What is the unit, and its abbreviation, that is used to measure electric charge?
                    page 35

6. Describe what an **electric current** is.
                    page 35

7. (a) What is the symbol for electric current?
    (b) What is the unit, and its abbreviation, that is used to measure electric current?
                    page 35

8. State the relationship between charge, current and time.   *page 35*

9. Calculate the current in a circuit when a charge of 180 C is transferred in 1 minute.

10. What is the difference between an **electrical conductor** and an **insulator**?   *page 35*

11. (a) Give *three* examples of electrical conductors.
    (b) Give *three* examples of electrical insulators.   *page 35*

Circuit symbols

12. Draw the circuit symbols for each of the following components:
    an ammeter; a voltmeter; a cell; a battery; a fuse; a switch; a lamp; a resistor; a variable resistor.
                    page 35

13. Identify the following circuit symbols.

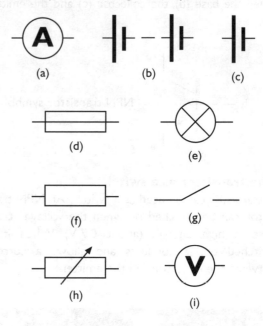

(a)   (b)   (c)

(d)   (e)

(f)   (g)

(h)   (i)

                    page 35

Voltage and potential difference

14. What is meant by the **voltage** of a supply?
                    page 35

15. (a) What is the symbol for voltage?
    (b) What is the unit, and its abbreviation, that is used to measure voltage?   *page 36*

16. Explain what is meant by the term **potential difference**.   *page 36*

17. Explain the similarities and the differences between the two terms 'voltage' and 'potential difference'.   *page 36*

18. What happens to the brightness of a lamp when the potential difference across it is increased?
                    page 35

19. Two quantities in electricity that are not the same but are often confused are voltage and current.
    Show that you understand these quantities by explaining what is meant by each of them.
                    page 36

Resistance

20. Explain what is meant by the term **resistance**.
                    page 36

21. (a) What is the symbol for resistance?
    (b) What is the unit, and its abbreviation, that is used to measure resistance?   *page 36*

Ammeters and voltmeters

22. What type of meter is used to measure current?   *page 36*

23. What type of meter is used to measure voltage?   *page 36*

24. Draw a circuit diagram to show how an ammeter is connected into a circuit.   *page 36*

25. Draw a circuit diagram to show how a voltmeter is connected into a circuit.   *page 36*

26. Redraw the following circuit to include a meter to measure the current through the resistor and a meter to measure the potential difference across the lamp. Use the correct circuit symbols for both meters.

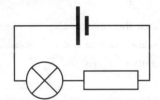

   page 36

27. What happens to the current in a circuit when the resistance in it is *increased*?   *page 36*

Series circuits

28. Describe what a **series circuit** is. Use a diagram to help your description.   *page 36*

29. Make a statement about the **current** at all positions in a series circuit.   *page 36*

30. State the relationship between the **voltage** of the supply and the potential differences across the components in a series circuit.   *page 36*

31. In the circuit shown, all of the resistors are identical. Copy the diagram and fill in the values on all of the meters, including units.

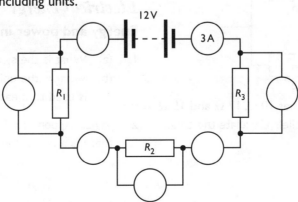

Parallel circuits

32. Describe what a **parallel circuit** is. Use a diagram to help your description.
   page 37

33. State the relationship between the **current** drawn from the supply and the currents in parallel branches in a parallel circuit.   *page 37*

34. Make a statement about the **potential difference** across components in a parallel circuit.   *page 37*

35. In the circuit shown, all of the bulbs are identical. Copy the diagram and fill in the values on all of the meters, including units.

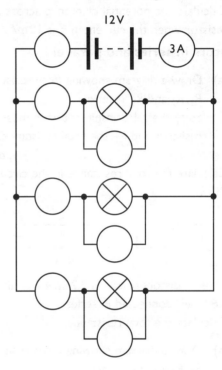

Resistance and Ohm's Law

36. State the relationship between resistance, current and voltage (potential difference) for a resistor. *page 37*

37. State **Ohm's Law** for a conductor. *page 37*

38. What is the ratio V/I for a resistor called? *page 37*

39. What happens to the ratio V/I for a resistor when the current in the resistor changes? *page 37*

40. Calculate the current taken from a 9V battery by a resistor of resistance $180\,\Omega$.

41. Calculate the potential difference across a $1\,k\Omega$ resistor when the current in it is $10\,mA$.

Resistors in series and parallel

42. (a) Draw a diagram showing *three* resistors R_1, R_2 and R_3 in series.
    (b) State the relationship between these resistors and R_T, the total resistance. *page 38*

43. Calculate the total resistance in the circuit shown.

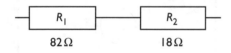

R_1 $82\,\Omega$ R_2 $18\,\Omega$

44. Three resistors of values $2.2\,\Omega$, $4.7\,\Omega$ and $5.6\,\Omega$ are connected in series. Calculate the total resistance.

45. (a) Draw a diagram showing three resistors R_1, R_2 and R_3 in parallel.
    (b) State the relationship between these resistors and R_T, the total resistance. *page 38*

46. Calculate the total resistance in the circuit shown.

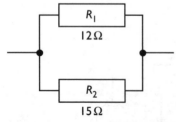

R_1 $12\,\Omega$

R_2 $15\,\Omega$

47. Three resistors of values $10\,\Omega$, $12\,\Omega$ and $15\,\Omega$ are connected in parallel. Calculate the total resistance.

The potential divider

48. Describe what a **potential divider** circuit is. *page 39*

49. Draw a potential divider circuit. *page 39*

50. Consider the potential divider circuit shown.

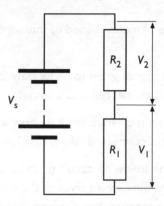

    (a) State the relationship between V_1, V_2, R_1 and R_2.

    (b) State the relationship between V_s, V_1 and V_2, and so give the relationship for V_1 in terms of V_s, R_1 and R_2. *page 39*

51. Calculate the potential difference V_1 in the potential divider circuit shown.

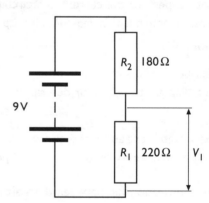

52. A potential divider consisting of two resistors is used to 'tap off' exactly half of a supply voltage. What is the relationship between the resistor values?

Electrical energy

Energy and power in electrical circuits

1. (a) What is the symbol for **energy**?
   (b) What is the unit, and its abbreviation, that is used to measure energy? *page 40*

2. What happens to the electrical energy when there is a current in a component? *page 40*

3. What happens to the brightness of a lamp when the current through it is increased?   *page 40*

4. State the relationship between:
   the rate at which electrical energy is transformed;
   voltage;
   current.   *page 40*

5. Explain what is meant by **power**.   *page 40*

6. (a) What is the symbol for power?
   (b) What is the unit, and its abbreviation, that is used to measure power?
   (c) How is the unit of power related to the joule?   *page 40*

7. A torch bulb takes a current of 0.25 A when operated from a 3 V battery.
   Calculate its power rating.

8. State the relationship between power, energy and time.   *page 40*

9. Calculate the current taken from the 230 V mains supply by a 460 W television.

10. Show how $V I$, $I^2 R$ and V^2 / R are all equal to each other and are all expressions used to find electrical power.   *page 40*

11. Calculate the amount of electrical energy transformed into other forms of energy every second in a 10 kΩ resistor when the current in it is 5 mA.

12. A car sidelight bulb is marked 5 W, 12 V. Calculate the resistance of its filament.

Fuses and appliances

13. (a) State the *two* sizes of fuse recommended for use in plugs.
    (b) Explain how to use the power rating to choose the correct fuse to use in the plug of a household appliance.   *page 41*

14. A floodlight lamp with a resistance of 92 Ω is connected to the 230 V mains supply.
    (a) Calculate the current drawn from the mains supply.
    (b) Use the calculated value of current to determine the fuse required for the lamp.

15. What is the job done by *all* electrical appliances?   *page 41*

16. What are the *two* forms of energy that electrical energy is transformed into in a lamp?   *page 41*

17. Describe the main energy transformation or transformations in each of the following household appliances:
    toaster, table lamp, kettle, light bulb, electric fire.   *page 41*

18. Name *three* appliances used in the home in which electrical energy is transformed into heat.   *page 41*

19. In which part of an electric heater does the energy transformation take place?   *page 41*

20. Explain why the heating element of an electric fire gets hot while the flex which connects the fire to the mains does not.   *page 41*

Direct current and alternating current

21. Explain what is meant by the term **direct current** (d.c.).   *page 42*

22. Explain what is meant by the term **alternating current** (a.c.).   *page 42*

23. What is the purpose of the **mains supply** or **battery** in an electrical circuit?   *page 42*

24. Which of the two types of electricity, a.c. or d.c., is supplied by:
    (i) the mains supply;
    (ii) a battery?   *page 42*

25. Draw the pattern which would be seen on a suitably adjusted oscilloscope when it is connected to:
    (i) a source obtained from the mains supply;
    (ii) a battery.   *page 42*

26. What is the **frequency** of the mains supply in Britain?   *page 42*

27. (a) Which is greater, the **quoted value** of an alternating voltage or the **peak value** of the same voltage?
    (b) Explain why there is a difference between the quoted value of an alternating voltage and the peak value of the same voltage.   *page 42*

28. Make a statement about the amount of power supplied to a given resistor by a d.c. supply and an a.c. supply of the same quoted value.   *page 42*

29. What is the **declared voltage** or the **quoted value** of the mains supply in Britain?   *page 42*

Electromagnetism

Magnets and magnetic fields

1. (a) What is a **magnetic field**?
   (b) How is a magnetic field usually represented? *page 42*

2. What effect does an electric current in a wire produce in the space around the wire? *page 42*

3. Draw the magnetic field around a straight current-carrying wire. *page 42*

4. Draw the magnetic field around a coil of wire carrying a current. *page 42*

Induced voltage

5. Replace the word 'induced' with another suitable word in the following statement:
   'A voltage was induced in a conductor.' *page 42*

6. Under what circumstances will a voltage be **induced** in a conductor? *page 42*

7. Draw *two* diagrams to illustrate two *different* situations where a voltage is induced in a conductor. *page 43*

8. There are *three* factors that affect the size of the voltage induced in a conductor.
   (a) State what these factors are.
   (b) How does each of them affect the size of the induced voltage? *page 43*

Transformers

9. What are **transformers** used for? *page 43*

10. Draw the transformer symbol. *page 43*

11. Describe the construction of a transformer. Use a labelled diagram to help your description. *page 43*

12. Explain how transformers can be used to change the size of an alternating voltage. Make sure your explanation includes why transformers only work with alternating voltages. *page 43*

13. Give the relationship between V_s, V_p, n_s and n_p for a transformer, saying what each of these terms means. *page 43*

14. (a) What is a 'step-up' transformer?
    (b) What is a 'step-down' transformer? *page 44*

15. A mains transformer used for a model train set has an output voltage of 11.5 V.
    Calculate the number of turns on the secondary winding if there are 2000 turns on the primary.

16. (a) Give an equation that can be used to calculate the **input power** to a transformer, defining each of the terms used in the equation.
    (b) Give an equation that can be used to calculate the **output power** from a transformer, defining each of the terms used in the equation. *page 44*

17. Give the relationship between the turns ratio and primary and secondary currents for an ideal transformer. *page 44*

18. An ideal transformer is used to allow an industrial machine designed to operate using a 115 V supply to be run from the mains supply of 230 V in Britain.
    (a) Is the transformer a step-up or a step-down transformer?
    (b) Calculate the turns ratio of the transformer.
    (c) When in use, the current taken from the secondary winding of the transformer is 12 A.
        Calculate the current in the primary winding.

Transmission lines and the National Grid

19. Describe how electrical energy is transmitted by the **National Grid** system. *page 45*

20. When electricity is transmitted long distances, power is lost.
    (a) Why is power lost?
    (b) State the equation that is used to calculate the power loss during transmission, defining all of the terms used in the equation.
    (c) State and explain how the power loss is kept as low as possible. *page 45*

21. Electricity is transmitted along transmission lines which have a resistance of 0.2 ohms per kilometre (Ω/km). The two transmission lines are each 25 km long.
    If the electricity is transmitted at a current of 20 A, calculate the total power lost in the transmission lines.

Electronic components

Input, process and output

1. What are the *three* main parts that an electronic system can be split into? *page 46*

2. Draw a **block diagram** showing how the three parts of an electronic system are linked.
   page 46

Output devices

3. What is an **output device** in an electronic system? *page 46*

4. (a) Give *four* examples of output devices.
   (b) For each of your examples, state the energy transformation involved. *page 46*

5. (a) State the main energy transformation in a **solenoid**.
   (b) Give a use for a solenoid. *page 46*

6. (a) State the main energy transformation in a **relay**.
   (b) Give a use for a relay. *page 46*

7. (a) State the main energy transformation in a **loudspeaker**.
   (b) Give a use for a loudspeaker. *page 46*

8. (a) State the main energy transformation in a **seven-segment display**.
   (b) Give a use for a seven-segment display.
   page 46

9. Explain how different numbers can be produced on a seven-segment display. *page 46*

10. What would be a suitable output device for each of the following situations?
    (a) To indicate that a battery powered CD player is switched on.
    (b) To display the time on a digital clock.
    (c) The sound system used by a pop group.
    (d) An electronically operated garage door bolt.
    (e) To switch on a high current, a.c. heater using a low current transistor circuit.

The LED

11. What does **LED** stand for? *page 47*

12. (a) Draw the symbol for an LED.
    (b) Indicate which way the electrons flow to make the LED emit light. *page 47*

13. Identify the following circuit symbol.

page 47

14. An LED is connected in a circuit and it emits light. The connections to the LED are then reversed. What happens? *page 47*

15. Why is a resistor needed in series with an LED?
    page 47

16. Draw a circuit diagram which will allow an LED to light.
    Include the series resistor and make sure the LED is connected the correct way to the supply.
    page 47

17. An LED and its series resistor are connected across a 9V battery.
    The potential difference across the LED is 2V and the current through it is 12.5mA.
    Calculate the value of the series resistor.

18. The filament lamp and the LED are both designed to transform electrical energy into light.
    Compare the electrical and the optical properties of both of these light emitters.
    page 47

Input devices

19. What is an **input device** in an electronic system? *page 47*

20. State the energy transformation that takes place in a **microphone**. *page 47*

21. State the energy transformation that takes place in a **thermocouple**. *page 47*

22. Give *one* use for a thermocouple. *page 47*

23. State the energy transformation that takes place in a **solar cell**. *page 47*

24. Give *two* uses for solar cells. *page 47*

25. Name *two* electrical components which have a resistance that changes due to a change in the physical conditions. *page 47*

26. Choose a suitable input device for each of the following applications.
    (a) Circuit to turn down the brightness of a television when the room lights are put out.
    (b) Alarm to warn parents in another room when a baby wakes up and cries.
    (c) Energy source for a satellite.
    (d) Temperature control for an aquarium.
    (e) To measure the temperature inside a furnace.

The thermistor and the LDR

27. What is a **thermistor**?   *page 48*

28. What usually happens to the resistance of a thermistor as its temperature increases?
                                         page 48

29. Draw the circuit symbol for a thermistor.
                                         page 48

30. Give *one* use for a thermistor.   *page 48*

31. What is an **LDR**?   *page 48*

32. What happens to the resistance of an LDR as the light intensity falling on it increases?
                                         page 48

33. Draw the circuit symbol for an LDR.   *page 48*

34. Give *two* uses for an LDR.   *page 48*

35. When the temperature of a thermistor alters, its resistance falls from 1000 Ω to 400 Ω. The thermistor has a potential difference of 5 V across it.
    Calculate the current in the thermistor at both temperatures.

36. An LDR has a potential difference of 2 V across it. As a result of a change in the light intensity falling on it, the current through the LDR changes from 10 mA to 25 mA.
    (a) Calculate the resistance of the LDR at both levels of light intensity.
    (b) Explain whether the light intensity falling on the LDR increases or decreases.

Transistors

37. Draw the circuit symbol for an n-channel enhancement MOSFET.   *page 48*

38. Explain where the name MOSFET comes from.
                                         page 48

39. Draw the circuit symbol for an NPN **transistor**.   *page 49*

40. Identify the following circuit symbols.

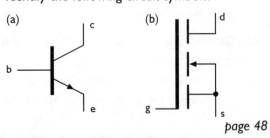

                                         page 48

The transistor as a switch

41. Give one use for a transistor.   *page 49*

42. A transistor has two possible states: on or off.
    Explain what is meant by this statement.
                                         page 49

43. Consider the circuit diagram shown below.

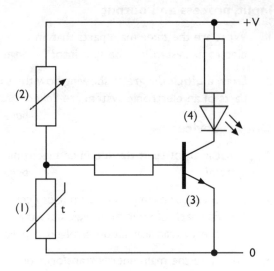

    (a) This circuit responds to a change in the environment.
        What is the quantity that changes, and in what way does the circuit respond to the change?
    (b) Explain how the circuit operates.
        In your explanation, name components (1), (2), (3) and (4) and state the purpose of each of them.

44. Consider the circuit diagram shown below.

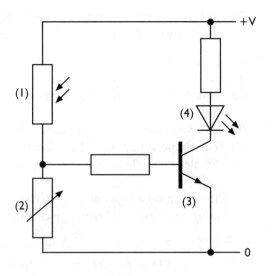

    (a) This circuit responds to a change in the environment.
        What is the quantity that changes, and in what way does the circuit respond to the change?
    (b) Explain how the circuit operates.
        In your explanation, name components (1), (2), (3) and (4) and state the purpose of each of them.

45. Consider the circuit diagram shown below.

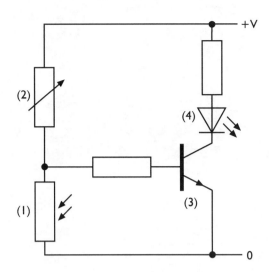

(a) This circuit responds to a change in the environment.
What is the quantity that changes, and in what way does the circuit respond to the change?

(b) Explain how the circuit operates.
In your explanation, name components (1), (2), (3) and (4) and state the purpose of each of them.

Amplifiers

46. What does an **amplifier** do in an electronic system? *page 49*

47. Which of the following electronic devices contain amplifiers?
radio; intercom; sewing machine; calculator; hi-fi system; model car circuit; personal stereo; baby alarm. *page 49*

48. How does the amplitude of the output signal from an audio amplifier compare with that of the input signal to the amplifier? *page 49*

49. How does the frequency of the output signal from an audio amplifier compare with that of the input signal to the amplifier? *page 49*

50. What is meant by the **voltage gain** of an amplifier? *page 49*

51. Give an equation that links input voltage, output voltage and voltage gain of an amplifier. *page 49*

52. Calculate the voltage gain of an amplifier which has an input voltage of 10mV and an output voltage of 0.5V.

53. Describe how to measure the voltage gain of an audio amplifier using an oscilloscope. *page 49*

WAVES AND OPTICS

Waves

Types of waves

A **wave** is a regular disturbance which carries energy and has no mass. Because waves transmit energy from one place to another, they can be used to send signals.

There are many types of wave. Some are carried by means of a medium, such as water waves. Others, like television and radio waves, can travel through a vacuum as well as through other substances like air. The waves that you should know about for this course are:

    water waves;

    sound waves;

    radio and television waves;

    light waves;

    other waves that together make up the electro-magnetic spectrum.

Radio and television signals travel through space as waves and transfer energy from place to place. The speed of radio and television signals through air and space is 300 million m/s (3×10^8 m/s).

Light also travels at a speed of 300 million m/s (3×10^8 m/s) in air or a vacuum.

Speed, distance and time

The relationship between distance, time and speed applies to all types of waves – water waves, sound waves, television and radio waves, microwaves and light waves.

$$\text{speed} = \frac{\text{distance}}{\text{time}} \qquad v = \frac{s}{t}$$

where v is speed in m/s

          s is distance in m

          t is the time in s.

EXAMPLE

Calculate how long it takes water waves to travel a distance of 15 m if the wave speed is 5 m/s.

SOLUTION

distance = 15 m

wave speed = 5 m/s

$$v = \frac{s}{t} \text{ so } t = \frac{s}{v} = \frac{15}{5} = 3$$

time = 3 s

Calculate how far light travels in one year.

SOLUTION

time = 1 year =
1 x 365 x 24 x 60 x 60s = 31 536 000 s
speed = 3 x 10⁸ m/s

$$s = v\,t = 3 \times 10^8 \times 31\,536\,000$$
$$= 9.46 \times 10^{15}$$

distance = 9.46 x 10¹⁵ m

The speed of sound

The **speed of sound** in air can be measured using the relationship between speed, distance and time:

$$\text{speed} = \frac{\text{distance}}{\text{time}} \qquad v = \frac{s}{t}$$

where v is speed in m/s
s is distance in m
t is time in s.

Because the speed of sound is so fast, when measuring it, it is necessary to use a timing device which can accurately measure a very short time period, or to use a very long distance.

Two methods which can be used to measure the speed of sound in air are the computer/interface method and the outdoor method.

Computer/interface method. Two microphones, a measured distance s apart (about 2 metres) are connected through an interface to a computer. The computer is used as a timing device to measure short time intervals. When a sound is made, it reaches microphone 1 first and this starts the computer timing. The same sound reaching microphone 2 stops the computer timing. The speed of sound is given by:

$$\text{speed of sound} = \frac{\text{measured distance } s}{\text{time recorded on computer } t}$$

Outdoor method. A buzzer and a light are switched on at exactly the same time a long, measured distance s from an observer. (The distance should be at least 50 metres to give a reliable result.) The observer starts a stopwatch when the light is seen and stops it when the sound from the buzzer is heard. The speed of sound is given by:

$$\text{speed of sound} = \frac{\text{measured distance } s}{\text{time recorded on stopwatch } t}$$

The approximate speed of sound in air is 340 m/s.

One example that illustrates the difference between the speeds of light and of sound in air is that lightning is seen before thunder is heard. Another example is at a fireworks display when the light from a firework is seen before the sound of it is heard.

Lightning in Tucson, Arizona

It is not necessary to use the speed of light in situations involving both a flash of light and a burst of sound because the speed of light is so great that the time for it to travel can be taken as instantaneous.

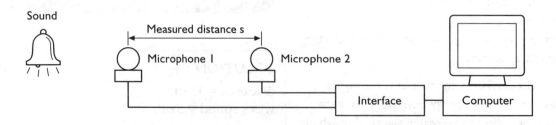

Measuring the speed of sound

EXAMPLE

A pupil uses two sound operated switches, 1.7 m apart, connected to an interface and a computer to obtain a value for the speed of sound. The computer records a time of 5 ms.

What value do these results give for the speed of sound?

SOLUTION

distance = 1.7 m
time = 5 ms = 0.005 s

$$v = \frac{s}{t} = \frac{1.7}{0.005} = 340$$

speed of sound = 340 m/s

Transverse and longitudinal waves

There are two types of wave: **transverse waves** and **longitudinal waves**. The difference is in the way the particles of the medium carrying the wave vibrate, in relation to the direction the energy is carried.

With a transverse wave, the particles of the medium vibrate at right angles to the direction in which the energy is carried (the direction of energy propagation). Examples of transverse waves include water waves, television and radio waves and light waves.

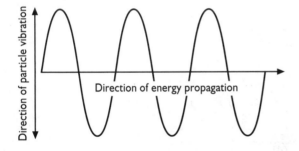

Sound is carried by longitudinal waves. When a sound is made in air for example, the air particles in front of the sound source are alternately pushed together and separated apart. These are known as compressions and rarefactions respectively of the air particles. It is this vibration of the air particles which carries the sound energy, along the same direction as the particle vibration.

A sound wave can be converted by a microphone into an electrical signal. This signal can be displayed on an oscilloscope as a transverse wave but the original sound that was made was carried as a longitudinal wave.

The **wavelength** of a wave is the distance from any point on one wave to the corresponding point on the next wave along. For example, from the top of one crest to the top of the next crest for a transverse wave or from one compression to the next compression for a longitudinal wave.

The symbol for wavelength is λ (Greek lambda). Since wavelength is a distance, it is measured in metres (m).

The **amplitude** of a transverse wave is the height of the wave from the centre position to the top of a crest, or the depth of the wave from the centre position to the bottom of a trough. The symbol for amplitude is *a*. Amplitude is measured in metres (m).

The following diagram of a section of a transverse wave shows what is meant by the terms wavelength, amplitude, crest and trough.

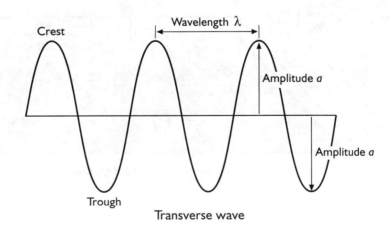

Transverse wave

The **frequency** of a wave is the number of waves that are made (or that pass a point) in a given time period.

The symbol for frequency is f. The unit used to measure frequency is the hertz (Hz) where one hertz is one wave or one cycle per second.

The **speed** of a wave is the distance the wave travels in a given time period. The symbol for wave speed is v. Speed is measured in metres per second (m/s).

The **period** of a wave is the time it takes to make one complete wave. The symbol for wave period is τ (Greek tau; sometimes T is used). The unit used to measure the period of a wave is the second (s).

The period and frequency of a wave are related as follows.

$$\tau = \frac{1}{f}$$

$$f = \frac{1}{\tau}$$

where τ is the wave period in seconds, s
f is the frequency in hertz, Hz.

The wave equation

The relationship between speed, frequency and wavelength for a wave is:

speed = frequency x wavelength

$$v = f\lambda$$

where v is speed in metres per second, m/s
f is frequencey in hertz, Hz
λ is wavelength in metres, m.

This relationship is known as the **wave equation**.

EXAMPLE

A wave generator in a swimming pool has a frequency of 10 Hz. The wavelength of the waves it produces is measured at 40 cm. Calculate the speed of the waves.

SOLUTION

frequency = 10 Hz
wavelength = 40 cm = 0.4 m
$v = f\lambda = 10 \times 0.4 = 4$
speed of waves = 4 m/s

Wave generator in a swimming pool

The electromagnetic spectrum

There are seven types of radiation which are known collectively as the **electromagnetic spectrum**. These are:

- gamma rays
- X-rays
- ultra violet
- visible light
- infra red
- microwaves
- television and radio.

All of these radiations travel at the same speed (3×10^8 m/s in a vacuum).

The wavelengths and frequencies of the radiations in the electromagnetic spectrum are different.

The approximate wavelength and frequency ranges for each of the radiations in the electromagnetic spectrum are as shown in the table.

Type of radiation	Approximate wavelength range (m)	Approximate frequency range (Hz)
gamma rays	10^{-13}–10^{-10}	10^{21}–10^{18}
X-rays	10^{-12}–10^{-8}	10^{20}–10^{16}
ultra violet	10^{-9}–10^{-7}	10^{17}–10^{15}
visible light	10^{-7}–10^{-6}	10^{15}–10^{14}
infra red	10^{-6}–10^{-3}	10^{14}–10^{11}
microwaves	10^{-4}–10	10^{12}–10^{7}
TV and radio	10^{-1}–10^{4}	10^{9}–10^{4}

EXAMPLE

Microwaves of wavelength 2.8 cm are produced by a microwave transmitter.
Calculate the frequency of the waves.

SOLUTION

wavelength = 2.8 cm = 2.8×10^{-2} m
speed = 3×10^8 m/s
(not stated explicitly)

$$f = \frac{v}{\lambda} = \frac{3 \times 10^8}{2.8 \times 10^{-2}} = 1.07 \times 10^{10}$$

frequency of waves = 1.07×10^{10} Hz

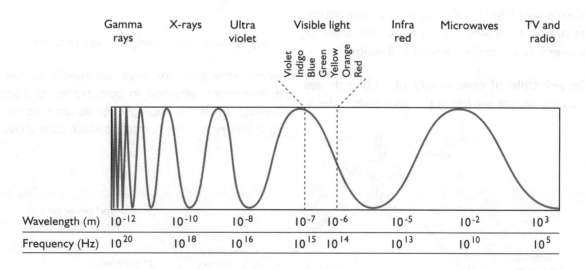

Gamma rays	X-rays	Ultra violet	Visible light	Infra red	Microwaves	TV and radio

Violet Indigo Blue Green Yellow Orange Red

| Wavelength (m) | 10^{-12} | 10^{-10} | 10^{-8} | 10^{-7} | 10^{-6} | 10^{-5} | 10^{-2} | 10^{3} |
| Frequency (Hz) | 10^{20} | 10^{18} | 10^{16} | 10^{15} | 10^{14} | 10^{13} | 10^{10} | 10^{5} |

The electromagnetic spectrum

Reflection

The law of reflection

Light travels in straight lines and is reflected when it meets a mirror.

There are three concepts that must be understood in relation to the reflection of a ray of light from a plane (flat) mirror. These concepts are **angle of incidence, angle of reflection** and **normal**. The diagram below illustrates these concepts.

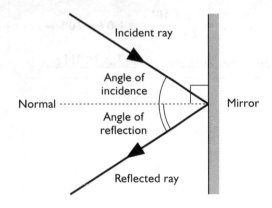

Reflection from a plane mirror

The normal is a line drawn at right angles (perpendicular) to the mirror surface at the point where the ray meets the mirror.

The angle of incidence is the angle between the incident ray and the normal drawn at the point where the ray meets the mirror.

The angle of reflection is the angle between the reflected ray and the normal drawn at the point where the ray is reflected off the mirror.

When a ray of light is reflected from a plane mirror, the angle of reflection is equal to the angle of incidence. This is the first **law of reflection**.

The **principle of reversibility** of a ray path says that a ray of light will follow the same path in the opposite direction when it is reversed. So if a ray of light is sent in the direction of the reflected ray in the diagram shown, it will in turn be reflected back along the direction of the original incident ray.

Curved reflectors

Curved reflectors attached to receiver aerials make the received signal stronger. The curved reflector attached to a receiver collects the signals from a larger area and reflects them to a focus. The aerial is positioned at the focus. The larger the reflector then the stronger is the signal directed to the aerial.

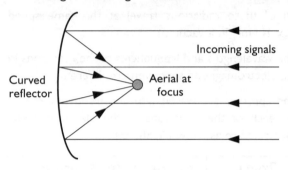

Signal reception using a curved reflector

Curved reflectors are used with certain transmitted signals to produce a parallel beam of signals for transmission. The reflectors do this because the transmitting aerial is placed at the focus of the curved reflector.

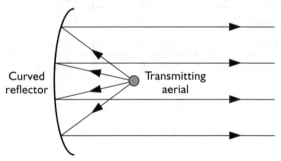

Signal transmission using a curved reflector

Curved reflectors are used extensively in tele-communication, attached to both transmitting and receiving aerials. For example, signals are sent to a geostationary satellite (a satellite which stays above

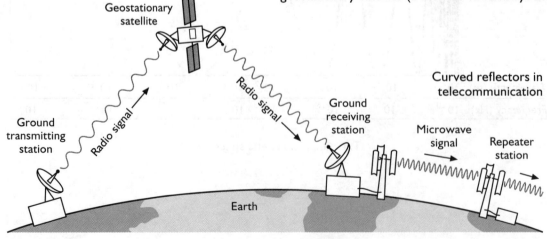

Curved reflectors in telecommunication

the same point on the Earth's surface) from a ground transmitting station using a curved reflector at the transmitting aerial. The satellite's receiving aerial is also at the focus of a small curved reflector.

The satellite then boosts or amplifies and re-transmits the signal to a ground station in the receiving country, again using a small curved reflector, this time associated with the transmitting aerial of the satellite. A group of three geostationary satellites is sufficient to allow communication right round the world from continent to continent.

The signals from the satellite are received at the aerial of a ground station which is also placed at the focus of a curved reflector. The ground station then transmits microwave signals close to the Earth's surface. Since microwaves can only travel a few kilometres they are sent to repeater stations where boosters amplify the signals and re-transmit them. The ground stations and the repeater stations all have curved reflectors at their receiving and transmitting aerials. This allows the signal strength to be kept as high as possible.

The critical angle and total internal reflection

When a ray of light travelling in glass meets a boundary with air, it can be **refracted** (see page 66) which means its direction usually changes. The angle of refraction in the air is greater than the angle of incidence in the glass. The angle of incidence in the glass which gives an angle of refraction of 90° is called the **critical angle** (usually about 42°). This is the greatest angle of incidence which allows re-fraction of light to take place.

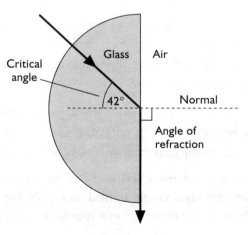

The critical angle

If a ray of light travelling in glass meets a boundary with air at an angle greater than the critical angle (about 42°), the ray is not refracted but reflected.

Since all of the ray is reflected inside the glass, this property is known as **total internal reflection**.

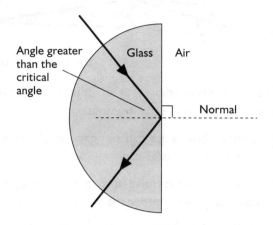

Total internal reflection

The value of the critical angle depends on the substances involved. For glass and perspex it is about 42° while the value for diamond is about 24°. This is one of the main reasons why a diamond sparkles so much — because with a properly cut diamond a greater amount of light is reflected out to the eye than with a gem that has a larger critical angle. When a diamond is viewed from above, a lot of light is seen which gives the impression of sparkling.

Diamond sparkling

Optical fibres

An **optical fibre** is a very thin, high-quality strand of glass which can carry light signals at very high speeds. The speed of light along an optical fibre is approximately 2×10^8 m/s. A bunch of optical fibres grouped together forms an optical cable which can carry a large number of signals at the same time between two places.

Light signals are transmitted along an optical fibre by using total internal reflection when they meet the wall of the fibre. This is because the angle of incidence when a light signal meets the wall of the optical fibre is greater than the critical angle in the glass.

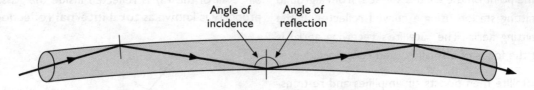

Angle of incidence   Angle of reflection

Light transmission along an optical fibre

There are many advantages of optical fibres over electrical cables for signal transmission.

- Optical fibre cables can be made much smaller than electrical cables can.

- Glass is less expensive than copper so optical fibre cables are cheaper than electrical cables.

- Optical fibre cables weigh less than electrical cables.

- Signals can be transmitted at very high speeds along optical fibre cables.

- Optical fibre cables can carry a lot more signals at the same time than electrical cables.

- The signals transmitted along optical fibre cables are light rays so they do not suffer from electrical interference. This means that the signal quality in an optical fibre cable is far better.

- Because less energy is lost in transmitting a signal along an optical fibre cable than along an electrical cable, fewer booster stations are required for the same transmission distance.

EXAMPLE

If an optical fibre communication link was to be laid between Buenos Aires and London it would be 11 000 km long.
Calculate how long it would take for a message to be transmitted along this link.

SOLUTION

distance = 11 000 km = 11 000 x 10³ m
speed = 2 x 10⁸ m/s

$$v = \frac{s}{t} \text{ so } t = \frac{s}{v} = \frac{11\,000 \times 10^3}{2 \times 10^8} = 0.055$$

__time = 0.055 s__

Refraction

Refraction of light

Refraction of light happens when light goes from one substance (or medium) into another, for example, from air into glass. When a ray of light is refracted, its direction usually changes.

There are three concepts that must be understood in relation to the refraction of light. These concepts are **angle of incidence**, **angle of refraction** and **normal**.

The diagram below illustrates these concepts.

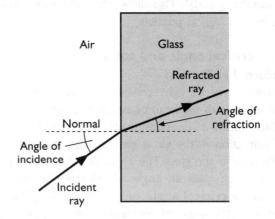

Refraction of light from air to glass

The normal is a line drawn at right angles (perpendicular) to the boundary between the two substances at the point where the ray goes from one substance to the other.

The angle of incidence is the angle between the incident ray and the normal, drawn at the point where the ray goes from one substance to the other.

The angle of refraction is the angle between the refracted ray and the normal, drawn at the point where the ray goes from one substance to the other.

When a ray of light passes from air into glass it usually changes direction and, in doing so, it bends to be closer to the normal. In this case, the angle of refraction is smaller than the angle of incidence.

A ray of light passing from glass into air will also usually change direction to be bent further away from the normal. In this case the angle of refraction is greater than the angle of incidence.

Both of these situations are shown in the diagram.

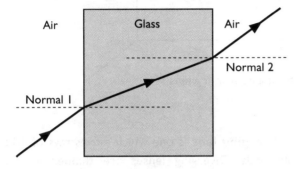

Refraction of light from air to glass
and back into air

Lenses

A **lens** is a shaped piece of glass or plastic which uses the property of refraction to change the direction of rays of light. There are two types of lens – **converging** and **diverging**.

A **converging lens** is one which brings rays of light closer together. Converging lenses are thicker in the centre than at the edges. They can be convex, plano-convex or even concave-convex, as the diagram shows.

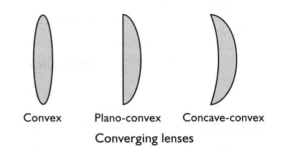

Converging lenses

Lenses are used in cameras, spectacles, binoculars, etc.

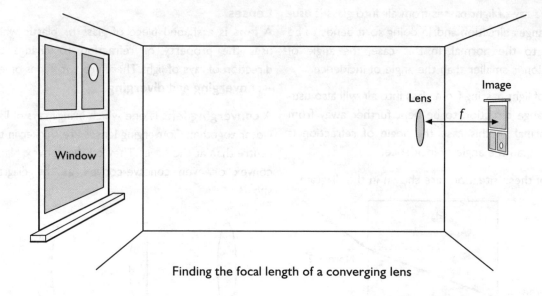

Finding the focal length of a converging lens

A converging lens bends parallel rays of light together to a focus as shown.

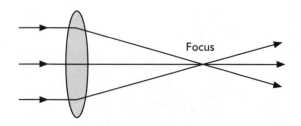

The effect of a converging lens

The **focal length** of a converging lens is the distance from the centre of the lens to the focus, where parallel rays of light from a distant object are focused by the lens into a sharp image.

To find the focal length of a spherical converging lens, use it to form a sharp image on one wall of the parallel rays of light coming through a window on the opposite wall. The focal length of the lens is the distance from the centre of the lens to the image on the wall.

A **diverging lens** is one which moves rays of light outwards. Diverging lenses are thinner in the centre than at the edges. They can be concave, plano-concave or even convex-concave, as the diagram shows.

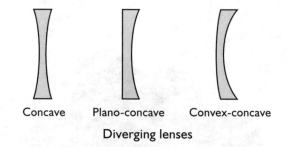

Diverging lenses

A diverging lens moves parallel rays of light outwards as shown.

Since a diverging lens moves parallel rays of light outwards, these rays can never cross. If the divergent rays are projected backwards they cross over at a point behind the lens. This point is called the **virtual focus** of the diverging lens and its distance away from the centre of the lens is the focal length of the lens.

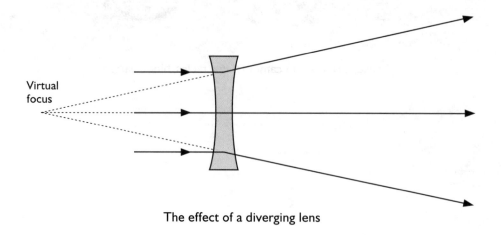

The effect of a diverging lens

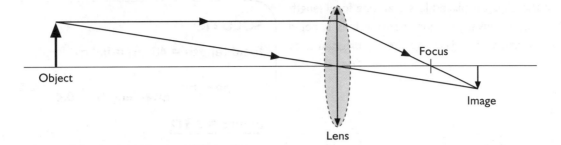

Ray diagram when object distance is greater than $2f$ from lens

Ray diagrams

When drawing ray diagrams, there are two rays coming from the object that are important in locating the image that the lens forms of the object:

- a ray that is parallel to the principal axis of the lens is sent through the focus of the lens;
- a ray that passes through the centre of the lens continues in the same direction.

Although refraction happens at both the curved surfaces of a lens, for simplicity it is more usual to consider that there is only one combined refraction which takes place in the central plane of the lens. For this reason the lens in a ray diagram is often represented by the symbol shown, the line being drawn in the central plane of the lens. In the ray diagrams which follow, this symbol is used with the lens itself shown dotted.

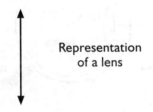

Representation of a lens

Different types of images are formed of an object by a converging lens, depending on the distance from the object to the lens.

The nature of the image formed is described in relation to three features or properties:

- **enlarged** or **diminished** (in relation to the object size);
- **upright** or **inverted**;
- **real** (can be formed on a screen) or **virtual** (cannot be formed on a screen).

When the object is placed at a distance which is more than two focal lengths in front of the lens, the ray diagram is as shown above.

The image formed is diminished, inverted and real.

The ray diagram above shows how the lens in a camera forms an image of the object on the photographic film. This is because the object is usually more than two focal lengths in front of the camera lens and so the real image produced on the film in the camera is smaller and upside-down.

When the object is between one and two focal lengths in front of the lens, the ray diagram is as shown below.

The image formed is enlarged, inverted and real.

Both a projector and an overhead projector use an object/lens configuration such as this to produce a real, enlarged image of an object like a slide.

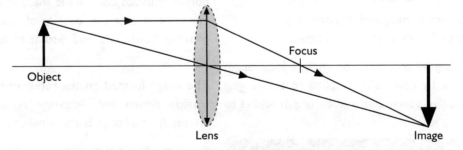

Ray diagram when object distance is greater than f and less than $2f$

When the object is placed less than one focal length in front of a converging lens, then the lens is being used as a magnifying glass and the ray diagram is as shown.

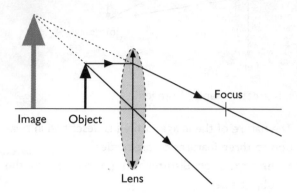

Image formed by a magnifying glass

Because the object is closer than one focal length from the lens, the two rays do not actually meet. When continued back, they do appear to meet. This defines where the image appears to be, as seen by the eye.

The image formed by a magnifying glass is:

* enlarged since it is larger than the object;
* upright (or erect) since it is the same way up as the object;
* virtual since the rays do not pass through where the image is seen.

Power of a lens

The **power** of a lens is a measure of the ability of the lens to bend light. The more curved a lens is then the higher is its power and the shorter is its focal length.

The power of a lens is measured in dioptres (D). The power and the focal length of a lens are related as follows.

$$power = \frac{1}{focal\ length}$$

where   power is measured in dioptres (D)
        focal length is measured in metres (m).

A negative sign when either a power or a focal length is quoted is, by convention, applied to a diverging lens. A positive power and/or focal length applies to a converging lens.

EXAMPLE

Calculate the power of a converging lens of focal length 40 cm.

SOLUTION

focal length = 40 cm = 0.4 m

$$power = \frac{1}{focal\ length} = \frac{1}{0.4} = 2.5$$

power = 2.5 D

EXAMPLE

Calculate the focal length of a diverging lens which has a power of −10 D.

SOLUTION

power = −10 D

$$focal\ length = \frac{1}{power} = \frac{1}{-10} = -0.1$$

focal length = −10 cm

The eye

The main parts of the **human eye** are as shown in the diagram.

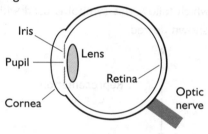

The human eye

Rays of light coming into the eye from an object are bent or refracted by both the cornea and the lens. The greater amount of bending happens at the cornea. Muscles control the shape of the lens to give fine control over the focusing of light on the retina, allowing both close and distant objects to be seen clearly.

The image formed on the retina of the eye is both **upside down** and **sideways reversed** (laterally inverted). The image is also smaller than the object.

A normal eye is able to focus on objects which are some distance away as well as objects which are close to it because the eye lens changes shape. When the eye is focusing on objects which are far away the lens is less curved. When the light comes from a closer object, the lens becomes more curved. This property of the eye is known as **accommodation**.

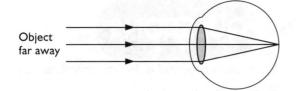

Object far away

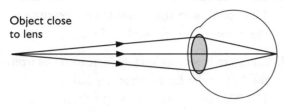

Object close to lens

Accommodation

Sight defects and their correction

A **long sighted** person can only see objects clearly which are a long distance away. The light from nearby objects is focused behind the retina.

A **short sighted** person can only see objects clearly which are close by. The light from far away objects is focused in front of the retina.

Long and short sight can be corrected by using lenses (either in spectacles or as contact lenses). The refraction of the light caused by the eye combined with the spectacle lens causes the light to be focused correctly on the retina of the eye.

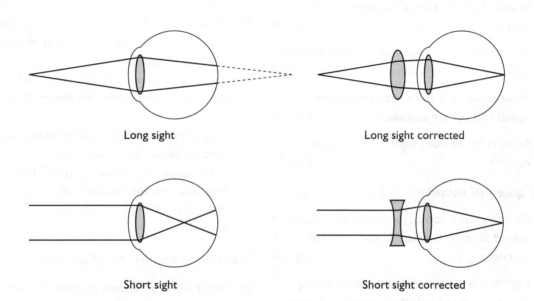

Long sight

Long sight corrected

Short sight

Short sight corrected

Waves

Types of waves

1. Explain what **waves** are.   *page 59*

2. Copy and complete:
   Because waves transmit _____ from one place to another, they can be used to send _____.
                                           page 59

3. **Sound** is carried by means of waves. Give at least three other types of waves.
                                           page 59

4. In what form do **radio and television signals** travel through space?   *page 59*

5. What do radio and television signals transfer from one place to another?   *page 59*

6. What is the speed of radio and television signals through air and space?   *page 59*

7. At what speed does **light** travel in air or a vacuum?   *page 59*

The speed of sound

8. Describe a method of measuring the **speed of sound** in air. A labelled diagram will help your description.   *page 60*

9. Explain why it is necessary to use a timing device which can accurately measure a very short time period (such as a microcomputer) or to use a very long distance when measuring the speed of sound.   *page 60*

10. What is the approximate speed of sound in air?   *page 60*

11. Give an example which illustrates the difference between the speed of light and the speed of sound in air.   *page 60*

Speed, distance and time

12. (a) State the relationship between **distance, time** and **speed**, using the symbols that are used for these quantities.

    (b) Give the units for each quantity.   *page 59*

13. Water waves move a distance of 10 m in 4 s. Calculate the wave speed.

14. Calculate how far water waves travel in 5 s if the wave speed is 3 m/s.

15. If the speed of sound is 340 m/s, calculate how far a sound will travel in 5 s.

16. Calculate how far away the Moon is if it takes 1.2 s for a radio signal sent from a beacon on the Moon to be received on Earth.

17. Calculate how long it takes light to travel from the Sun to the Earth when the Sun is 150 000 000 km away from the Earth.

18. Explain why it is not necessary to use the speed of light in problems involving both a flash of light and a burst of sound.   *page 60*

19. During a thunderstorm, a girl notices that the sound of thunder comes 3 s after she has seen the flash of lightning.
    If the speed of sound in air is 340 m/s, calculate how far away from the storm the girl is.

20. A lighthouse sends out a flash of light and a burst of sound at the same time.
    If the speed of sound in air is 340 m/s, how long after the light is seen will an observer on the bridge of a ship 1.36 km away hear the sound?

Transverse and longitudinal waves

21. What is the difference between a **transverse wave** and a **longitudinal wave**?   *page 61*

22. (a) Give an example of a transverse wave.
    (b) Give an example of a longitudinal wave.
                                           page 61

23. (a) What is meant by the **frequency** of a wave?
    (b) What is the symbol for frequency?
    (c) What is the unit used to measure frequency?   *page 62*

24. (a) What is meant by the **wavelength** of a wave?
    (b) What is the symbol for wavelength?
    (c) What is the unit used to measure wavelength?   *page 61*

25. (a) What is meant by the **speed** or **velocity** of a wave?
    (b) What is the symbol for wave speed?
    (c) What is the unit used to measure speed?
                                           page 62

26. (a) What is meant by the **amplitude** of a transverse wave?
    (b) What is the symbol for amplitude?
    (c) What is the unit used to measure amplitude? *page 61*

27. (a) What is meant by the **period** of a wave?
    (b) What is the symbol for period?
    (c) What is the unit used to measure period? *page 62*

28. (a) Draw a diagram of a transverse wave.
    (b) Mark the following on your diagram:
        a crest;
        a trough;
        one wavelength;
        the amplitude. *page 61*

29. What is the relationship between wave period τ and wave frequency f? *page 62*

30. What energy change takes place in a **microphone**? *page 61*

31. What is the name of the instrument that can be used to look at the patterns of waves produced by a microphone? *page 61*

32. What is meant by the **frequency** of a sound? *page 62*

33. What is meant by the **amplitude** of a sound? *page 61*

34. Copy and complete the following diagrams to show the signal patterns obtained when the sound signals are as stated.

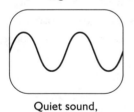

Quiet sound, low frequency

Quiet sound, high frequency

Loud sound, low frequency

Loud sound, high frequency

The wave equation

35. (a) State the relationship between **speed**, **frequency** and **wavelength**, using the correct symbols for these quantities.
    (b) Give the units for each quantity. *page 62*

36. A tuning fork produces a note of 262 Hz. Calculate the wavelength of the waves produced in air when the wave speed is 340 m/s. Use an appropriate number of figures in your answer to this question.

37. Calculate how many times a boat will bob up and down a second if it is in a harbour with waves of wavelength 50 cm travelling at 25 cm/s.

The electromagnetic spectrum

38. (a) Name the seven types of radiation which collectively are known as the **electromagnetic spectrum**.
    (b) What do all of these radiations have in common?
    (c) What properties are different for the radiations in this family of waves? *page 63*

39. Copy and complete the following table to include all seven types of radiation in the electromagnetic spectrum.

Type of radiation	Approximate wavelength range (m)	Approximate frequency range (Hz)
	shortest	highest
	longest	lowest

page 63

40. A solid state laser emits light which has a wavelength of 660 nm (1 nanometre = 1×10^{-9} m). Calculate the frequency of the light emitted.

41. BBC Radio 5 Live broadcasts on a frequency of 909 kHz.
    Calculate the wavelength of the radio waves transmitted.

42. The wavelength of the radio waves transmitted by the Max AM radio transmitter is 194 m. Calculate the frequency allocated to the Max AM radio station in kHz.

Reflection

The law of reflection

1. How do rays of light travel? *page 64*

2. What happens to a ray of light when it meets a mirror? *page 64*

3. Explain what is meant by the terms **angle of incidence, angle of reflection** and **normal** in relation to a ray of light being reflected from a plane mirror. *page 64*

4. Draw a diagram showing a ray of light being reflected from a plane (flat) mirror. Label the angle of incidence, the angle of reflection and the normal on your diagram. *page 64*

5. Describe the direction of the ray of light reflected from a flat mirror, in relation to the direction of the incident ray. *page 64*

6. State the first **law of reflection** for a ray of light. *page 64*

7. The paths of rays of light show the **principle of reversibility**.
   What does this principle state? *page 64*

Curved reflectors

8. What is the purpose of curved reflectors on some receiver aerials? *page 64*

9. Use a diagram to explain why curved reflectors are used with some received signals. *page 64*

10. What is the purpose of curved reflectors on some transmitter aerials? *page 64*

11. Use a diagram to explain why curved reflectors are used with some transmitted signals. *page 64*

12. Name four types of signals which can be focused by curved reflectors.

13. Describe an application of curved reflectors used in telecommunication. In your description you should mention:
    satellite communication;
    microwave or TV links;
    repeater stations;
    boosters. *page 64*

The critical angle and total internal reflection

14. (a) Explain what is meant by total internal reflection.

    (b) Draw a diagram to show total internal reflection of a ray of light. *page 65*

15. (a) Explain what is meant by the **critical angle**.

    (b) Draw a diagram to show the critical angle for a ray of light in a glass block. *page 65*

Optical fibres

16. What is an **optical fibre**? *page 65*

17. By what means are signals transmitted along an optical fibre? *page 65*

18. Copy and complete a larger version of the diagram below to show how a ray of light travels from one end of an optical fibre to the other. Mark the angles of incidence and reflection on your diagram.

page 66

19. What is the approximate speed of signals which are transmitted along an optical fibre? *page 65*

20. Describe the principle of operation of an optical fibre transmission system. *page 65*

21. Give one practical example of signal transmission which uses optical fibres.

22. Both electrical cables and optical fibres are used for signal transmission.
    Optical fibres have at least *seven* advantages over electrical cables for signal transmission.
    What are these advantages? *page 66*

Refraction

Refraction of light

1. What is meant by **refraction** of light? *page 66*

2. Draw a diagram to show the path of a ray of light which passes from air into glass. *page 66*

3. Draw a diagram to show the path of a ray of light which passes from glass into air. *page 66*

4. (a) Draw a diagram showing a ray of light being refracted.
    Include the normal in your diagram.

    (b) Label the angle of incidence and the angle of refraction on your diagram. *page 66*

5. Copy and complete the diagram to show the path of the ray of light as it enters and leaves the glass block.

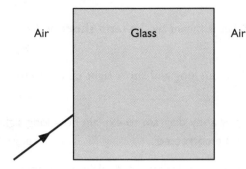

Air   Glass   Air

Include and label on your diagram all rays, angles and relevant normals. *page 67*

Lenses

6. Describe the shape of a **converging (convex) lens**.
   Draw a diagram to help your description.
   page 67

7. Describe the shape of a **diverging (concave) lens**.
   Draw a diagram to help your description.
   page 68

8. Draw a diagram to show what a converging lens does to parallel rays of light passing through it.
   page 68

9. Draw a diagram to show what a diverging lens does to parallel rays of light passing through it.
   page 68

10. One type of astronomical telescope is a **refracting** telescope. Explain where this name comes from.

11. What is meant by the **focal length** of a converging lens? *page 68*

12. Describe a simple experiment to find the focal length of a converging lens. *page 68*

Ray diagrams

13. (a) When drawing ray diagrams, what two rays coming from the object are important?
    (b) Why are these rays important in locating the image that the lens forms of the object?
    page 68

14. What three features have to be stated when describing the image formed by a converging lens? *page 69*

15. (a) Draw a ray diagram to show how the converging lens of a camera forms the image of an object when it is at a distance of more than two focal lengths from the lens.
    (b) Describe the nature of the image formed by the lens. *page 69*

16. (a) Draw a ray diagram to show how the converging lens of a projector forms the image of a slide when it is at a distance of between one and two focal lengths from the lens.
    . (b) Describe the nature of the image formed by the lens. *page 69*

17. Explain what a **magnifying glass** is. *page 70*

18. (a) Draw a ray diagram to show how an image is formed by a magnifying glass.
    (b) Describe the nature of the image formed by a magnifying glass. *page 70*

19. Use a ray diagram to show how a real, diminished and inverted image is formed on the retina of the eye by the cornea and lens in the eye. *page 69*

Power of a lens

20. What is meant by the **power** of a lens?
    page 70

21. What is the unit that is used to measure the power of a lens? *page 70*

22. State the equation that links the power and the focal length of a lens. *page 70*

23. A lens is quoted as having a power of −4 D. What is the significance of the negative sign?
    page 70

24. Calculate the power of a converging lens of focal length 25 cm.

25. Calculate the focal length of a diverging lens which has a power of −20 D.

The eye

26. Draw a simple diagram of an **eye** and label the following parts:
    the iris and the pupil;
    the cornea and the lens;
    the retina and the optic nerve. *page 70*

27. Describe how light is focused on the retina of the eye. *page 70*

28. What differences are there between an object and the image of it that is formed on the retina of the eye? *page 70*

29. The eye is usually able to focus on objects which are some distance away from it as well as objects which are close to it. This is called **accommodation**.
How is the eye able to do this? *page 70*

30. Draw a ray diagram to show how a normal eye forms an image of an object which is some distance from the eye. *page 71*

31. Draw a ray diagram to show how a normal eye forms an image of an object which is close to the eye. *page 71*

Sight defects and their correction

32. What is meant by the term **long sight**? *page 71*

33. What is meant by the term **short sight**? *page 71*

34. How can long and short sight be corrected? *page 71*

35. Draw a ray diagram to explain how long sight can be corrected. *page 71*

36. Draw a ray diagram to explain how short sight can be corrected. *page 71*

RADIOACTIVITY

Ionising radiations

The structure of the atom

An atom consists of a **nucleus** containing **protons** and **neutrons**, surrounded by orbiting **electrons**.

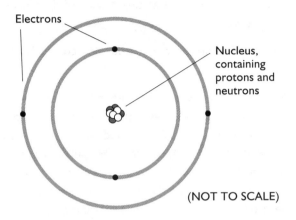

Protons and neutrons are approximately equal in mass. The mass of an electron is about $\frac{1}{2000}$ that of a proton. So most of the mass of an atom is concentrated in the nucleus.

- A proton carries a **positive** charge.
- An electron carries a **negative** charge.
- A neutron has **no charge**.

The **mass number** of an atomic nucleus is the number of protons and neutrons in the nucleus.

Types of radiation

When the nucleus of an atom disintegrates and forms the nucleus of a different type of atom, it gives out **radiation.** When radiation passes through a medium, the energy it carries may be absorbed by the medium.

The three types of radiation are **alpha particles** (α), **beta particles** (β) and **gamma rays** (γ).

Although not strictly a type of radiation, moving neutrons like those present in the nuclear fission process (*see page 84*) have energy and also exhibit some of the properties of radiation. Slow moving neutrons are sometimes called **thermal neutrons.**

The range through air of each type of radiation is as follows.
- Alpha radiation can only travel a few centimetres in air.
- Beta radiation can travel through a few metres of air.
- Air does not absorb gamma radiation.

The minimum thickness of different materials needed to absorb each type of radiation is as follows.

- Alpha radiation is absorbed by thin paper.
- A few millimetres of aluminium will absorb beta radiation.
- Gamma radiation is absorbed by a minimum of a few centimetres of lead.
- Gamma radiation is best at penetrating the human body.

Alpha radiation consists of positively charged particles. An alpha particle is a helium nucleus. It is made up of two protons and two neutrons, so it has a mass number of 4 and a charge of +2.

Many radioactive materials can emit alpha radiation. When uranium-238 decays (or changes) to thorium-234, an alpha particle is emitted from the uranium-238 nucleus.

Beta radiation consists of fast moving electrons. The mass of an electron is about $\frac{1}{2000}$ that of a proton. So the mass number of beta radiation is 0. Beta radiation has a charge of −1.

There are many radioactive materials that emit beta radiation. For example thorium-234 decays to protactinium-234 by emitting beta particles, one from each nucleus that decays.

Gamma radiation is a burst of electromagnetic radiation of very short wavelength and high energy.

When an atom decays by emitting either alpha or beta particles, the nucleus that remains is often unstable. It can become more stable by rearranging itself and in the process it emits a burst of gamma radiation. This happens, for example, with the thorium-234 formed from the decay of uranium-238.

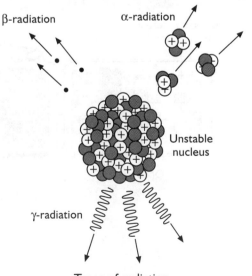

Types of radiation

Ionisation

An atom is normally uncharged because it contains equal numbers of protons and electrons, and therefore equal numbers of positive and negative charges.

When an atom is **ionised** it gains electrons to become negatively charged or loses electrons to become positively charged. The charged particle is known as an **ion**.

Alpha particles and beta particles can ionise atoms which they hit by knocking electrons away from the atoms. For this reason these types of radiation are known as **ionising radiations**.

Alpha particles produce much greater ionisation density than beta particles or gamma rays because alpha particles are the largest and they carry the greatest charge of all types of radiation. Because of their large size, alpha particles are the least penetrating and so their concentration is greatest on the surface of a body. Also because of their large size and the fact that their charge is greatest, alpha particles ionise the greatest number of atoms near the surface of a body.

In a **smoke detector**, a small sample of the radioactive material americium-241 is used to ionise air particles. These ions are made to flow as an electric current. When there is smoke present, the smoke particles become attached to the ions and this reduces the size of the current. This in turn sets off an audible alarm as a warning.

Effects of radiation on non-living things

Three examples of the effects of radiation on non-living things are:

- radiation **causes ionisation**;
- radiation **fogs photographic film**;
- radiation **causes scintillations**.

When radiation hits certain materials, the energy it contains is re-emitted as tiny flashes of light, called scintillations.

These effects are used in various detectors of radiation.

A **Geiger–Müller** tube contains a low pressure gas. When any radiation (even alpha) enters the thin mica window it ionises this gas and allows a current between the two electrodes. This pulse of current is recorded on a scaler or a counter connected to the tube. This means that a Geiger–Müller tube can be used as a radiation detector giving a count rate for various radioactive sources.

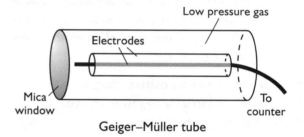

Geiger–Müller tube

A **radiation film badge** contains a small piece of photographic film behind various thicknesses of different absorbers of radiation. When the film is developed, the amount of fogging at each part of it gives an indication of the amount and the intensity of any radiation that the badge and hence the wearer have been exposed to.

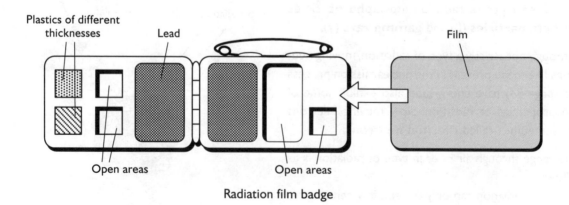

Radiation film badge

The scintillations caused by radiation are made use of in **scintillation counters** and **gamma cameras**. In the early days of research into radiation, these scintillations were individually counted by scientists such as Geiger and Marsden. Nowadays some patients are given drugs combined with chemicals which emit gamma radiations. These radiations are detected by a gamma camera which builds up images of internal organs. The gamma radiations cause scintillations when they reach crystals in the camera and these flashes of light in turn produce electrical impulses which are used to build up images of the organs.

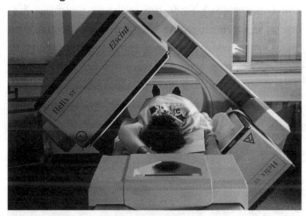

Gamma camera

Effects of radiation on living cells

Radiation can kill living cells or change their nature. Two medical uses of radiation based on the fact that radiation can destroy cells are:

- to **sterilise instruments** (by killing bacteria);
- to treat cancer, a process called **radiotherapy**.

Medical instruments such as scalpels and knives, as well as pre-packaged plastic syringes and needles which are intended to be used only once and then disposed of, are **sterilised** by exposing them to radiation. This radiation kills off all bacteria so the instruments do not spread infections when used.

In **radiotherapy**, radiation from a source is directed to the cancer cells in a body.

In one method, a cobalt-60 source is rotated around the patient. The gamma rays that the source emits can be concentrated at the cancerous tissue while causing less damage to the healthy surrounding tissue.

Another method of treatment involves the patient swallowing a meal containing a radioactive substance which concentrates in a particular organ needing treatment. For example, cancer of the thyroid gland can be treated with the beta radiation emitted from iodine-131.

Tracers

There are various uses of radiation in medicine, agriculture and industry, based on the fact that radiation is easy to detect. In most of these applications, the radioactive material is used in the form of a **tracer**.

In medicine, a tracer is a radioactive substance which is taken into the body either as an injection or as a drink (sometimes called a 'barium meal'). The gamma radiation which it gives off is monitored and gives an indication of any problems there may be in an organ by how much (or how little) of the substance is absorbed by the organ.

In agriculture, it is important to know how well plants make use of fertilisers. To do this, a small amount of the radioactive substance phosphorus-32 is sometimes added to the fertiliser. Its progress through the plant can then be monitored, to give an indication of how the fertiliser is being utilised by the plant.

In industry, radioactive substances can be used to monitor the flow of liquids. Leaks in underground water and sewage pipes can be detected by adding radioactive tracers to the liquids in the pipes and monitoring the radioactivity in the soil surrounding them. Similarly, by the use of tracers, pollution levels or water flow in rivers can be monitored, using mobile detectors.

Dosimetry

The activity of a source

The rate at which a radioactive source decays is called the **activity** of the source. It is the number of nuclei which disintegrate in a given time period.

The symbol for activity is A. The activity of a radioactive source is measured in **becquerels (Bq)**. One becquerel is defined as one nucleus decaying per second.

One becquerel is a very small unit of activity. Some approximate values of the activity of various radioactive sources are given in the table.

Source	Activity in Bq
1 litre of seawater	10
school radioactive source	2×10^5
sources used in medicine	1×10^9
1 gram radium	3×10^{10}

Absorbed dose

When a radioactive source decays, it gives out energy. This energy can be absorbed by another material and can cause damage to the material. The **absorbed dose** is the energy absorbed per unit mass of the absorbing material.

The symbol for absorbed dose is D. The unit for absorbed dose is the **gray (Gy)** where one gray is one joule per kilogram (J/kg).

When the energy given out by all types of radiation is absorbed by living tissue, it can cause damage to the tissue. The amount of damage depends on both the source and the tissue.

The risk of biological harm from an exposure to radiation depends upon three factors:
* the absorbed dose;
* the kind of radiation that is absorbed, for example α, β, γ or slow neutrons;
* the type of tissue or body organ that absorbs the radiation.

Radiation weighting factor

Different types of radiation have different effects on living cells. Even although the same type of tissue may receive the same dose, the biological effects of different radiations will be different. To take this into account, a radiation weighting factor is assigned to all types of radiation. The **radiation weighting factor** allows the ability of different types of radiation to damage living cells to be compared.

The symbol for the radiation weighting factor is w_R. The radiation weighting factor for radiation does not have a unit because it is not a physical quantity. It is a number which gives the relative harm done to cells by different types of radiation.

The radiation weighting factor for various types of radiation is shown in the table.

Radiation	Radiation weighting factor (w_R)
alpha particles (α)	20
beta particles (β)	1
gamma rays (γ)	1
slow (thermal) neutrons	3

Equivalent dose

Radiation damage in biological systems depends on the type of radiation as well as how the energy that it releases is distributed. The **equivalent dose** is a measure of the biological effect of radiation and it takes account of the type and energy of the radiation as well as how the radiation is distributed.

The symbol for equivalent dose is H. Equivalent dose H is the product of absorbed dose D and radiation weighting factor w_R.

$$H = D\, w_R$$

The unit that is used to measure equivalent dose is the **sievert (Sv)**. Since the radiation weighting factor w_R is a dimensionless constant, both the gray (the unit of absorbed dose) and the sievert (the unit of equivalent dose) are equal to one joule per kilogram. The sievert is introduced to assess potential biological damage from radiation as well as to distinguish equivalent dose from absorbed dose.

EXAMPLE

A worker in the nuclear industry received an absorbed dose of 400 μGy from slow neutrons. Calculate the equivalent dose received.

SOLUTION

absorbed dose $D = 400\ \mu Gy = 400 \times 10^{-6}$ Gy
radiation weighting factor w_R for slow neutrons = 3 (not stated explicitly)
equivalent dose H = ?

$$H = D\, w_R$$
$$H = 400 \times 10^{-6} \times 3 = 1.2 \times 10^{-3}$$

equivalent dose received = 1.2 mSv

Worker in the nuclear industry

Background radiation

We are all constantly subjected to both natural and man-made radiation called **background radiation**. Except for people who are being treated in hospital by radiation and also some workers in the nuclear industry, the largest part of the radiation that most people are subjected to comes from natural sources. When a radiation detector is used in the absence of any obvious radioactive sources, it will record a count.

Natural sources of background radiation include the following.

Cosmic radiation which arrives on Earth from both the Sun and outer space. At sea level this contributes an annual equivalent dose of around 0.3 mSv but it increases with altitude, by about 20% for 1000 m height. A flight in a jet aircraft at its usual cruising height can increase the equivalent dose rate to about 200 times that at sea level.

The Earth. Soil and rocks in some areas are naturally radioactive and therefore the building materials used in construction are also radioactive. The granite found in the Aberdeen area and the rocks of Cornwall are both high in radioactivity. This source can contribute as much as cosmic radiation, about 0.3 mSv per year.

The air. The radioactive gas radon occurs naturally in the air, from rocks and building materials and is usually dispersed. However in poorly ventilated buildings, radon can build up to dangerously high levels.

The human body. Our bodies contain potassium, a percentage of which is the radioactive form potassium-40. We also eat food which has been grown in the soil and is therefore slightly radioactive. The average equivalent dose per year is about 0.4 mSv.

Man-made sources which contribute to background radiation include the following.

Medical sources. X-rays and cancer treatment in hospitals are the main sources of man-made radiation. Chest X-rays can be the major contributors to equivalent dose received per year – about 2 mSv.

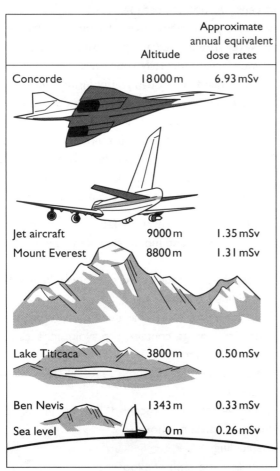

	Altitude	Approximate annual equivalent dose rates
Concorde	18000 m	6.93 mSv
Jet aircraft	9000 m	1.35 mSv
Mount Everest	8800 m	1.31 mSv
Lake Titicaca	3800 m	0.50 mSv
Ben Nevis	1343 m	0.33 mSv
Sea level	0 m	0.26 mSv

Cosmic radiation and altitude

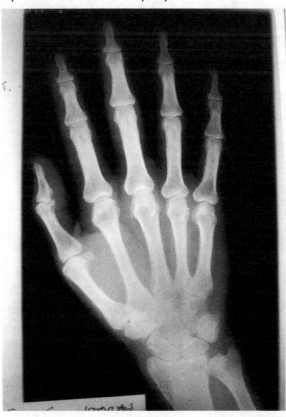

X-ray

Industry. Nuclear reactors used to generate electricity, and industry which uses radioactive sources mainly for monitoring purposes, both contribute about 1 mSv per year to equivalent dose received.

Half-life and safety

Measuring half-life

The activity of a radioactive source decreases with time.

However the rate of decrease slows with time. Because of this, and because the decay of individual atoms is random and unpredictable, theoretically a radioactive source will never completely lose all of its activity.

The time taken for half of the atoms in a radioactive sample to decay is a constant for that source called the **half-life** of the source. So the half-life of a radioactive source is the time period during which the activity of the source falls to half of its original value.

The half-life of some sources is as low as a fraction of a second; for others it is many thousands of years.

To measure the half-life of a radioactive source, the level of the background radiation is first measured. Then the count rate with the radioactive source present is measured over an appropriate period of time using a suitable detector such as a Geiger–Müller tube connected to a scaler. A graph of the count rate (with the source present), corrected for background radiation, is plotted.

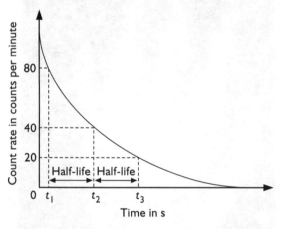

A suitable count rate value is chosen, say 80 counts per minute, and the time at which the source had this count rate, t_1, is marked as above. In a similar way the time t_2 at which the count rate is half the previous value, 40 counts per minute, is found. The half-life of the source is the time period $t_2 - t_1$. *Any* starting value can be chosen, the time period for the count rate to halve in value will always be the same.

EXAMPLE

In six years, the activity of a radioactive isotope drops from 200 kBq to 25 kBq.
Calculate the half-life of the isotope.

SOLUTION

original activity = 200 kBq

activity after 1 half-life = $\frac{1}{2}$ x 200 kBq = 100 kBq

activity after 2 half-lives = $\frac{1}{2}$ x 100 kBq = 50 kBq

activity after 3 half-lives = $\frac{1}{2}$ x 50 kBq = 25 kBq

So 6 years represents 3 half-lives,

<u>thus one half-life is 2 years.</u>

Safety with radiation

There are several safety precautions that must be taken when handling radioactive substances.

- Always handle radioactive substances with forceps. Do not use bare hands.
- Never point radioactive substances at anyone.
- Never bring radioactive substances close to your face, particularly your eyes.
- Wash hands thoroughly after using radioactive substances – especially after using open sources or radioactive rock samples.
- Unauthorised people must not be allowed to handle radioactive substances. In particular, in the United Kingdom, no one under 16 years of age may handle radioactive substances.

In addition there are several safety precautions relating to the storage and monitoring of radioactive substances.

- Always store radioactive substances in suitable lead-lined containers.
- As soon as a source has been used, return it to its safe storage container, to avoid unnecessary contamination.
- Keep a record of the use of all radioactive sources.

The equivalent dose received by people can be reduced by three methods:

- shielding;
- limiting the time of exposure;
- increasing the distance from the source.

Radiation shielding door

Shielding is done in several ways. For example doctors, dentists and radiotherapists using radioactive sources wear lead-lined aprons or go behind lead shields and view through windows with lead in the glass. Hospitals often site their radioactive sources underground and behind concrete shields, away from busy areas.

Limiting the time of exposure. When dentists take an X-ray photograph of a patient's teeth, they will ask the patient to hold the X-ray film to limit the overall time that the dentist is exposed to radiation – remember dentists are exposed to a lot more X-rays than one patient is.

Increasing the distance. This is perhaps the simplest way of reducing the exposure to the radiation from a source. The intensity of radiation falls off according to the square of the distance from the source. In industry, workers usually handle sources by remote control using tongs to keep them as far away from the sources as possible.

This radioactivity warning sign is used where radioactive materials are stored or in use. The colours of the sign are black on a yellow background.

RADIOACTIVE MATERIALS

Nuclear reactors

Nuclear power

A **power station** is where electricity is generated using stored energy which is contained either in a fuel or in some other source.

A **thermal** power station burns a fossil fuel (coal, gas or oil), to release as heat the energy that is stored in the fossil fuel.

A **nuclear** power station uses the energy released when a radioactive element decays into other elements. The most common fuel that is used in a nuclear reactor is **uranium**.

Sellafield nuclear power station, Cumbria

The three main parts of a nuclear power station are shown in the block diagram below.

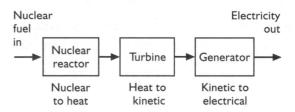

There are many advantages and disadvantages of using nuclear power for the generation of electricity. It is important to take account of both sides of the discussion for and against nuclear power, since an advantage can be countered by an equally valid disadvantage.

The **advantages** claimed for nuclear power include the following.
* It is cheaper than electricity produced by other fuels.
* Nuclear power does not use fossil fuels, which are non-renewable sources of energy and are in limited supply.
* Small amounts of waste materials are produced for large amounts of nuclear energy produced.

- Nuclear power does not contribute to the greenhouse effect whereas burning fossil fuels does.
- Fast breeder nuclear reactors can be used to produce additional fuel, plutonium, for use in reactors.

Amongst the **disadvantages** of nuclear power are the following.

- There is a risk of an accident at a nuclear reactor endangering large numbers of people.
- The uranium used in nuclear reactors has to be mined and then refined – both of which contribute to the unseen hazards.
- The world's supply of uranium will in fact run out earlier than the supply of fossil fuels.
- The plutonium which is produced in fast breeder nuclear reactors is the material from which nuclear bombs are made.
- Nuclear reactors produce dangerous radioactive waste which has to be carefully stored and controlled for many years.

Nuclear fission

Energy is obtained from the nuclear fuel uranium by a process called **nuclear fission**. Nuclear fission is the process of breaking up the nucleus of an atom into smaller nuclei. In the case of the uranium used as fuel in a nuclear reactor, this fission is caused by the bombardment of the uranium nuclei by neutrons. Such a process is called **induced fission**.

It is also possible for **spontaneous fission** to occur; this usually happens with larger atomic nuclei.

Fission occurs as follows in a nuclear reactor.

Neutrons are used to bombard the nuclei of uranium atoms. When a neutron hits the nucleus of a uranium-235 atom, it is absorbed and the nucleus becomes unstable (Diagram 1).

If there is sufficient energy, the nucleus then splits into two smaller nuclei and up to three neutrons are also released. Because the fragments gain kinetic energy during this process, energy in the form of heat is produced (Diagram 2).

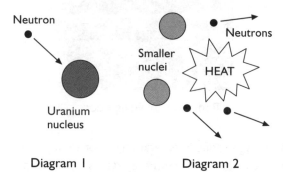

Diagram 1   Diagram 2

When the fission of a uranium-235 nucleus takes place, one of the products of the fission reaction is up to three neutrons. If the concentration of uranium-235 nuclei is high enough, each of these neutrons can go on to cause fission of further uranium-235 nuclei. As long as the conditions remain favourable, the neutrons produced continue to cause fission of further nuclei in ever increasing numbers. This type of reaction is called a **chain reaction**. The diagram below shows three stages of such a chain reaction.

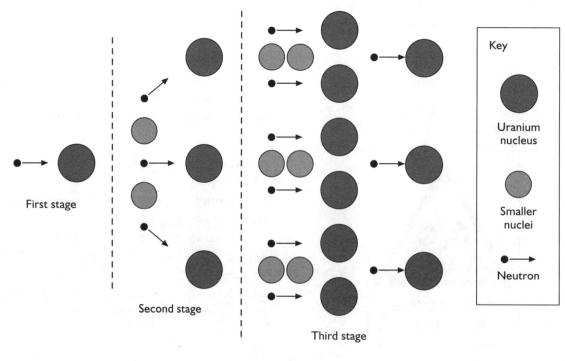

Chain reaction

The nuclear reactor

The **core** of a nuclear reactor is the part where the nuclear fuel is used up and the heat is produced. The essential parts of the core are:

- the **fuel** in the form of rods;
- a **moderator** to slow down the neutrons;
- **control rods** to control the rate of the reaction;
- a **coolant** to take away the heat from the core.
- the **containment vessel** which can absorb the gamma radiation produced and also withstand the heat and the pressure.

The **fuel** usually used is enriched uranium dioxide which is made in the form of pellets. These pellets are stacked together to form fuel rods, which are in turn bunched together in groups of 36 to form fuel elements.

The neutrons which are released during fission travel very fast, too fast to go on reliably to cause further fission. To sustain the reaction, the neutrons are slowed down by making them collide with the nuclei of another material in the core – the **moderator**.

The moderator is usually carbon in the form of graphite, although water or heavy water can also be used. The moderator does not absorb the neutrons, it slows them down by successive collisions, the neutrons eventually becoming thermal neutrons.

If all of the thermal neutrons were allowed to cause further fission, the result would be an uncontrolled chain reaction, generating greater and greater amounts of heat. To prevent this, **control rods** made of boron or carbon are used. These rods are raised and lowered automatically in response to changes of temperature, sensed by thermistors in the core of the reactor. The material of the control rods absorbs the excess neutrons and so prevents them from causing further fission. This in turn regulates the heat output of the core.

A further set of control rods is normally suspended above the core of the reactor and these are ready to fall into the reactor core in an emergency, if the core becomes too hot. When all of the control rods are inserted into the core, the reaction is shut down completely.

Fast breeder nuclear reactor

The heat produced by the nuclear reaction is transferred to the **coolant** – a liquid or gas, often carbon dioxide under pressure – and this in turn boils water to produce steam. The steam is then used to turn a turbine which generates electricity as in a conventional fossil fuel power station.

The core of a nuclear reactor is surrounded by a **containment vessel** which must be able to protect operators from gamma rays and neutrons as well as withstand the high temperatures and pressures of the reactor in normal use. This containment vessel is usually made of thick concrete lined in steel.

A secondary function of the containment vessel is that it must be able to contain all of the contents of the core in the event of an emergency.

One of the major difficulties with the use of nuclear power is the radioactive waste that it produces. The waste products can be divided into three categories:

   (i)   low radioactivity solid waste;

   (ii) medium level radioactive waste;

   (iii) low volume highly radioactive waste.

The first two categories of waste provide no more major storage problems than the waste from other industries except that, being radioactive, it is subject to government regulations.

The highly radioactive waste can remain radioactive for thousands of years and so has to be carefully stored and disposed of. At present it is encapsulated in drums which are then overpacked in cases. These cases are then stored in vaults in deep, stable geological formations.

In future the high level radioactive waste may be incorporated in glass blocks which are resistant to heat, radiation and moisture.

RADIOACTIVITY QUESTIONS

Ionising radiations

The structure of the atom

1. Describe a simple model of an **atom**. Include a labelled diagram to help your description. *page 77*

2. (a) Name the three main particles that make up an atom.
   (b) What are the relative masses of each of the particles that make up an atom? *page 77*

3. What type of charge does each of the following particles carry: proton, electron, neutron? *page 77*

Types of radiation

4. What happens to the energy that radiation carries when it passes through a body? *page 77*

5. Name the *three* types of radiation. *page 77*

6. State the range in air of all three types of radiation. *page 77*

7. State the minimum amount of material that will absorb each of the three types of radiation. *page 77*

8. Which of the three types of radiation is best at penetrating the human body? *page 77*

9. Explain what an **alpha particle** is, giving its mass number and the type of charge it carries. *page 77*

10. Explain what a **beta particle** is, giving its mass number and the type of charge it carries. *page 77*

11. Explain what **gamma rays** are. *page 77*

12. State what causes gamma radiation to be emitted from an atom. *page 77*

Ionisation

13. Why is an atom normally uncharged even although it contains charged particles? *page 78*

14. When radiation passes through a material it can **ionise** the atoms of that material. Explain what is meant by the term ionisation. *page 78*

15. Explain briefly how a **smoke detector** operates. *page 78*

16. Which of the three types of radiation produces the greatest ionisation density of atoms? *page 78*

Effects of radiation on non-living things

17. (a) Give *three* examples of the effects of radiation on non-living things.
    (b) For each of the three examples, describe how the effect is used in a detector of radiation, naming the detector in each case. *page 78*

Effects of radiation on living cells

18. State *two* effects that radiation can have on living cells. *page 79*

19. Describe *two* medical uses of radiation based on the fact that radiation can destroy cells. *page 79*

Tracers

20. Explain what is meant by a **tracer**. *page 79*

21. Describe *one* use of radiation based on the fact that radiation is easy to detect, in each of the three areas of medicine, agriculture and industry. *page 79*

Dosimetry

The activity of a source

1. Explain what is meant by the **activity** of a radioactive source. *page 79*

2. What is the symbol for the activity of a radioactive source? *page 79*

3. (a) What unit is used to measure the activity of a radioactive source?
   (b) How is this unit defined? *page 79*

Absorbed dose

4. State and explain what is meant by **absorbed dose**. *page 80*

5. What is the symbol for absorbed dose? *page 80*

6. (a) What is the unit of absorbed dose?
   (b) How is this unit defined? *page 80*

7. The risk of biological harm from an exposure to radiation depends upon three factors. What are they? *page 80*

Radiation weighting factor

8. Explain what is meant by the radiation weighting factor for radiation. *page 80*

9. (a) What is the symbol for the radiation weighting factor?
   (b) Why does the radiation weighting factor for radiation not have a unit? *page 80*

10. Copy and complete the following table, inserting the radiation weighting factor for each of the radiations given.

Radiation	Radiation weighting factor (w_R)
alpha particles (α) beta particles (β) gamma rays (γ) slow (thermal) neutrons	

page 80

Equivalent dose

11. Explain what is meant by **equivalent dose**. *page 80*

12. What is the symbol for equivalent dose? *page 80*

13. Give the relationship between equivalent dose, absorbed dose and the radiation weighting factor. *page 80*

14. What is the unit that is used to measure equivalent dose? *page 80*

15. What is the significance of equivalent dose? *page 80*

16. A person receives an absorbed dose of 40 µGy from a radioactive source which emits alpha particles only.
    Calculate the equivalent dose received by the person.

(The radiation weighting factor for alpha particles is 20.)

17. Calculate the equivalent dose received in a year by a worker in the nuclear industry who is subjected to the following emissions during the year.
    200 µGy from slow neutrons of radiation weighting factor 3
    50 µGy from gamma radiation of radiation weighting factor 1

Background radiation

18. What is meant by **background radiation**? *page 81*

19. (a) State three sources that contribute to the background radiation level.
    (b) For each source that you have given, indicate approximately how much it contributes to the background radiation level. *page 81*

Half-life and safety

Measuring half-life

1. What happens to the activity of a radioactive source as time goes on? *page 82*

2. What does the term **half-life** mean? *page 82*

3. Explain why it is necessary to define the activity of a radioactive source in terms of half-life. *page 82*

4. Describe the principles of a method of measuring the half-life of a radioactive source. *page 82*

5. A radioactive source has a half-life of 15 days. Calculate its activity 60 days after it was measured at 1600 kBq.

6. A sample of radioactive iodine-131 in a medical physics laboratory had its activity monitored at the same time every week for nine weeks. The background radiation count over the same

Week	1	2	3	4	5	6	7	8	9
Recorded activity in counts/minutes	140.0	82.0	50.0	33.0	23.5	18.0	15.5	14.0	13.0
Corrected activity in counts/ minute									

period was monitored and found to be a constant 12 counts per minute.

(a) Copy and complete the table, giving the corrected count rate for the sample each week.

(b) Plot a graph of the corrected count rate for the sample during the period shown in the table.

(c) Use your graph to estimate the half-life of iodine-131.

Safety with radiation

7. Describe *some* safety precautions which it is necessary to take when handling and storing radioactive substances. *page 82*

8. State the *three* ways that could be used to reduce the equivalent dose received by a person. *page 82*

9. Draw the radioactivity warning sign and state where this sign is used. *page 83*

Nuclear reactors

Nuclear power

1. What is a **power station**? *page 83*

2. Some power stations are known as **thermal** power stations.
Why thermal? *page 83*

3. What is the most common fuel that is used in a nuclear reactor? *page 83*

4. Give *some* of the means that are used to generate electricity other than nuclear power. *page 83*

5. State the advantages of using nuclear power over other means for the generation of electricity. *page 83*

6. State the disadvantages of using nuclear power over other means for the generation of electricity. *page 84*

Nuclear fission

7. Explain what is meant by **nuclear fission**. *page 84*

8. Nuclear fission can be either spontaneous or induced.
Explain the difference between these two processes. *page 84*

9. Describe in simple terms how the process of fission is used in a nuclear reactor to produce heat. *page 84*

10. Explain in simple terms what is meant by a **chain reaction**.
A diagram may help your explanation. *page 84*

The nuclear reactor

11. Describe the principles of the operation of a nuclear reactor.
Your answer should include a description of and the function of each of the following parts of the nuclear reactor:
(i) the fuel rods (pins);
(ii) the moderator;
(iii) the control rods;
(iv) the coolant;
(v) the containment vessel. *page 85*

12. Describe the problems associated with the storage and disposal of radioactive waste products from nuclear reactors. *page 86*

HOW PRACTICAL ACTIVITIES ARE ASSESSED

Since physics is a practical subject, you will be involved in doing experimental work throughout the Course. At some stage in the Course, one of your experimental reports will be formally assessed to see how well you are doing in the practical activities. This assessment should not be treated as a 'special event' but just taken as part of the practical work that is carried out in class. You may well carry out the experiment as part of a group, although the report must be your own work.

There is no particular experiment which *must* be carried out and assessed but it is likely that the experiment that is chosen by your teacher for assessment will come from the following list. If your teacher selects a different experiment it must cover some of the work from the Course and be of a suitable standard for assessment.

Mechanics and Heat

1. How the instantaneous speed of an accelerating object changes with time.
2. Measuring the acceleration of a trolley released at different places on a slope.
3. Measuring the acceleration of a trolley on different angles of slope.
4. How the acceleration of an object changes as the unbalanced force is changed.
5. How the acceleration of an object changes as its mass is changed.
6. Investigating cooling curves.

Electricity and Electronics

1. How the current through a resistor changes with potential difference.
2. How the resistance of a lamp changes with the current through it.
3. The relationship between peak voltage and the quoted value of voltage.
4. How the resistance of a thermistor changes as its temperature changes.
5. The relationship between V_{out} and V_{in} for an amplifier.

Waves and Optics

1. Measuring the speed of sound in air.
2. How the angle of refraction varies with the angle of incidence.
3. Measuring the focal length and calculating the power of various lenses.

Radioactivity

1. Finding the half-life of a radioactive source.
2. How the count rate from a beta source changes with the thickness of an aluminium shield.
3. How the count rate from a gamma source changes with the thickness of a lead shield.

(Because of safety regulations, it is possible that the information for your radioactivity report may come from viewing a video.)

When you have to do one of these experiments, it is likely that you will be given an instruction sheet to work from. Sample instruction sheets for all of the suggested experiments are on pages 103–119. While the instruction sheet will contain all the necessary information for you to carry out the experiment safely, there are certain things that it will not contain – you have to work these out for yourself.

- The heading will tell you the aim or objective of the experiment but will not give away the result. For example 'How the acceleration of an object changes as the unbalanced force is changed' not 'To show that acceleration is proportional to unbalanced force'.

- A diagram or a circuit diagram of the apparatus to use is likely to be supplied.

- You will be given instructions about how to carry out the experiment and what measurements to make but you will not be given a blank table or format for a table – you will have to decide the best form to use to record your results. You will also have to mention what instruments you used to make each of the measurements.

- The instruction sheet will probably not tell you to repeat your measurements but, where appropriate, you should do so.

- After you have taken down your results, you will have to analyse them. Once again, you will not be specifically told how to do this. The words that will probably be used will be 'Use an appropriate format to show ...'. What this means is that you should either plot a graph of your results or calculate further values and enter them in an expansion of your table of results.

- You will also be expected to come to a valid and relevant conclusion about your experiment and to evaluate the experimental procedures that were used (backing this up with reasons). It is unlikely that the instruction sheet will mention either of these two parts of your report.

Once again, this is starting to look very daunting. However, you will probably also be given a 'help sheet' of advice which you can refer to while you are doing the experiment and also while you are writing your report. The help sheet gives general guidelines on the style of English used for a scientific report and how to structure a report. It is important to follow this structure since that is the basis on which the report is marked. You will not be given a pro-forma report sheet which has the structure and format of the report built in.

A version of the help sheet is given here.

Advice on doing a scientific experiment and writing a report

When writing a scientific report, it is best to write 'The distance between the microphones was measured.' not 'I measured the distance between the microphones.' This is known as the passive impersonal voice in the past tense. Using the wrong voice will not lead to your report being returned but is bad practice, so should be avoided.

Avoid irrelevant details, such as 'The apparatus was collected from the cupboard.'

Keep the description simple and keep your sentences short.

Organise or structure your report in a logical order, usually in the order in which the experiment was carried out, using the headings given here.

Title: Usually this can be copied from the instruction sheet.

Aim or objective: A simple statement saying what the purpose or reason of the experiment is.

Apparatus: This can either be a list (possibly copied from the instruction sheet) or a neat line drawing which is clearly labelled. Remember, the clearer the better.

Procedure: A brief description which includes:
- what the independent variable was (that is the quantity that was changed in the experiment);
- what instrument was used to measure the independent variable and how it was changed;
- how all other measurements were taken and/or observations were made (remember to include the instrument used).

Results: Readings or observations are recorded in a clear table which:
- has clear headings giving the quantity being measured, with the units, as appropriate;
- has all of the results entered correctly, in a logical order;
- gives repeated measurements, where appropriate.

Analysis: The readings or observations are analysed in an appropriate way – either as a table or as a graph.
If a table is used, it can be either extra columns added to the table of results or it can be a separate table. In either case there must be correct headings, correct units and the calculations used must be shown.
If a graph is used, the two variables, independent and dependent, must be on the appropriate axes (independent variable almost always on the x-axis). Both axes must be labelled with the variable quantity and units and also have suitable scales making the best use of the graph paper. The best-fitting line (straight or curved) indicated by the points must also be drawn.

Conclusion: There must be a valid conclusion (otherwise there is no point in doing an experiment). This would include a statement about any of the following which are appropriate:
- an overall pattern to the readings or observations;
- a trend in the information or results that has been analysed;
- a connection that can be observed between the quantities;
- the measurement of a physical quantity.

Evaluation: The procedures used in the experiment must be evaluated and supported by a reason. A good evaluation could contain any of the following points:
- the effectiveness of the procedures that were used in the experiment;
- possible sources of error;
- how variables were or could have been controlled;
- improvements that could have been made to the experiment;
- limitations which were caused by the equipment.

When your report is completed it will be marked by your teacher and your work assessed. If you have followed the advice on the help sheet, it is likely to be of an acceptable standard.

Your teacher will check that your work has met the following criteria.

(a) You have been actively involved in collecting the information for the experiment.
(This will have been checked as you did the experiment.)

(b) You have described the experimental procedures accurately.
(This information comes from the *aim, apparatus* and *procedure* sections of your report.)

(c) The information and observations that you have made are relevant and suitably recorded.
(From your table of *results*.)

(d) All your results have been correctly analysed and shown in a suitable way.
(The *analysis* section of your report shows this.)

(e) You have reached a valid conclusion from your results.
(The *conclusion* section of your report.)

(f) You have evaluated the experimental procedures and supported this evaluation with a reason or reasons.
(The last section of your report, the *evaluation*.)

You may find, however, that your teacher considers that some or all of your work does not reach the standard needed. In this case you will be given your report back to amend and resubmit. You will not be told directly what else is needed but it will be marked with a cross ($\times$) where something is wrong or an omission mark ($\wedge$) where you have missed something out. You then have the chance to alter or add to your report and then resubmit it.

The four examples of marked student experimental reports which follow show some of the more common errors in writing reports which would lead to them having to be modified and resubmitted. The first report is acceptable in all areas. The comments added to these reports are to help you – any of your reports which have to be resubmitted will not have these helpful comments added.

Investigating cooling curves

Aim

To investigate the difference in cooling curves of water and stearic acid

Apparatus

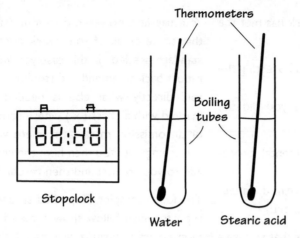

Thermometers

Boiling tubes

Stopclock

Water   Stearic acid

List:
Boiling tube with stearic acid
Boiling tube with water
2 thermometers
Water bath
Test tube rack
Stopclock

Procedure

The tubes of water and acid were heated in a water bath to 80 °C. The independent variable in this experiment was time which was measured using a stopclock. Every minute the temperature of the water and stearic acid was measured using thermometers.

Results

Time in minutes	Temperature of water in °C	Temperature of stearic acid °C
0	80	80
1	74	71
2	72	69
3	68	68
4	67	68
5	65	68
6	62	68
7	60	68
8	59	68
9	57	68
10	55	67
11	53	67
12	52	67
13	51	67
14	49	67
15	48	67
16	47	67
17	45	67
18	44	67
19	44	67
20	43	67

Analysis

See graph.

Conclusion From doing this experiment, it has been found that the temperature of stearic acid drops quicker than water over the first minute of the experiment but after 3 minutes the temperature of the stearic acid stays constant at around 68–67 °C. The stearic acid also begins to solidify.

The water drops in temperature in a more stable way, dropping by 1–2 °C after every minute. From this experiment it can be said that stearic acid retains a high temperature for longer than water.

Evaluation Possible sources of error are that both substances may not have been exactly 80 °C when the stopclock was started. It was hard to take both the temperatures at exactly the same time. An assistant to take one reading would help. Limitations caused by the equipment were that the thermometers were only accurate to 1 °C.

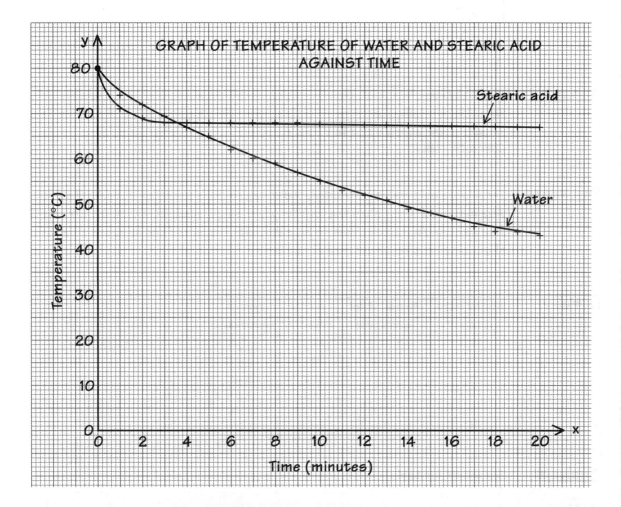

Teacher's comments and decision

This report contains all the information that is necessary to meet all of the criteria, so it is acceptable in all areas.

The only comment that might be made about it is that both the conclusion and the evaluation are rather wordy. While it is better to be concise in these areas, it is not essential.

How the resistance of a lamp changes with the current through it

__Aim__   To investigate how the resistance of the lamp changes with current through it.

Circuit diagram

Circuit diagram

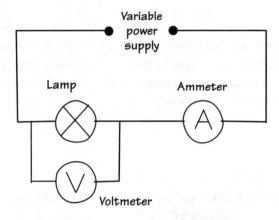

__Apparatus__   12 V, 24 W lamp
Ammeter and voltmeter
Variable power supply up to 12 V at 2 A
Connecting leads

__Procedure__   The above circuit was constructed, the voltage across the lamp ✗ and the current were measured ✗ for when the variable power supply read: 2V, 4V, 6V. 8V, 10V and 12V. This was repeated three times to help eliminate errors and the averages were taken. I then worked out the resistance of the lamp.

__Results__

Nominal output voltage (V)	Average voltage across lamp ✗	Average current (A)	Average resistance ✗
2	✗ 0.49	✗ 0.41	1.2
4	1.79	0.69	2.59
6	3.2	0.92	3.48
8	4.7	1.12	4.2
10	6.25	1.29	4.84
12	7.46	1.42	5.25

__Analysis__   See graph and table.

__Evaluation__   A possible source of error was that the power supply used did not give very accurate output voltages. I tried to combat this by repeating the experiment three times and using the averages.
Also when trying to read the voltage across the lamp on the digital voltmeter it was hard to be accurate. ✗

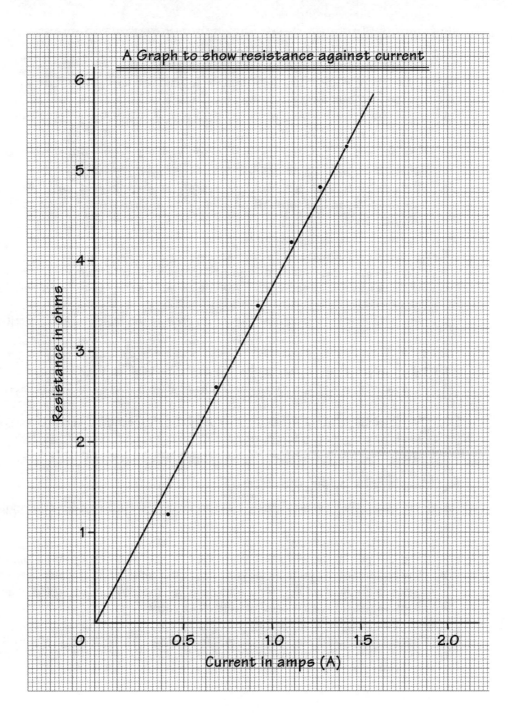

A Graph to show resistance against current

Resistance in ohms

Current in amps (A)

Criterion (b) – experimental procedure

1. No problems with the aim or the apparatus list.
2. There is no need to label standard circuit symbols on the circuit diagram – not a reason for resubmission, however.
3. It is more correct to use 'potential difference' across the lamp not 'voltage' – also not a reason for resubmission, however.
4. The procedure changes from third person to first person – bad practice, but again not a reason for resubmission.
5. The procedure does not mention that potential difference was measured using the voltmeter and current measured using the ammeter – this makes the report fail criterion (b).

Criterion (c) – recorded observations

1. The procedure mentions that repeated measurements were made but only the averages have been recorded on the results table – this makes the report fail criterion (c).
2. No unit for voltage (potential difference) given in the table – again fails criterion (c).

Criterion (d) – analysis of results

1. The table does not show how the values of R were calculated – fails criterion (d).
2. No units for resistance given and 'amps' is not an acceptable abbreviation for current – again fails criterion (d).
3. The best-fit line has not been drawn – again fails criterion (d).

Criterion (e) – conclusion

Since no conclusion has been given, this report obviously fails to meet criterion (e).

Criterion (f) – evaluation

1. The first part of the evaluation is not really valid – at best it might be ignored.
2. The second part is valid but does not contain a supporting argument – so it fails criterion (f). It could be made acceptable by the addition of a statement like 'because the reading changed between two values'.

So, although this report is quite well written, it does not meet any of the necessary criteria.

How the angle of refraction varies with the angle of incidence

The aim of the experiment is to see if there is a relationship between the angle that light enters a perspex block and the angle that it leaves.

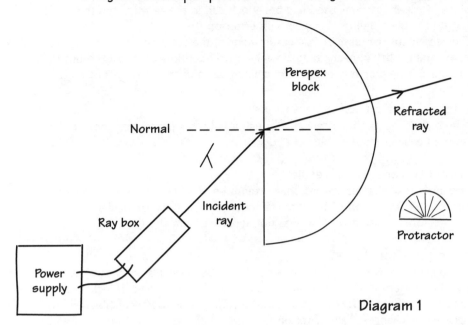

Diagram 1

The independent variable was the angle of incidence. The angle started at 10° and ended on 70°.

A measurement of the refracted ray ✕ was taken every 10° (10°, 20°, 30°, etc.) To do this the incident rays and refracted rays were marked on a piece of paper.

A protractor was used to measure all the angles.

Results

Angle of incidence	Angle of refraction
10°	7°
20°	13°
30°	19°
40°	26°
50°	31°
60°	36°
70°	40°

Conclusion

From the table of results it can be said that as the angle of incidence changes, the angle of refraction also changes ✕ but there is no obvious relationship.

Evaluation

Possible sources of errors are not doing the experiment in a dark enough room so not being able to take the readings properly; the perspex block moving so there is no constant error (random). Improvements could have been fixing the block down, doing the experiment in a darker room and doing it more than once.

Criterion (b) – experimental procedure

1. There are no problems with the aim or the labelled diagram of the experiment.
2. Although the angle of incidence is mentioned, it is not defined. Perhaps if the piece of paper that was used had been included in the report then this angle might have been defined – report fails criterion (b).
3. The statement that says '. . . a measurement of the refracted ray was taken . . .' is vague and probably means that the angle of refraction was measured but since this angle is neither mentioned nor defined, again criterion (b) is not met.

Criterion (c) – recorded observations

Here the report meets the criterion although it is better practice to include the units in the headings of the table.

Criterion (d) – analysis of results

Since no analysis of the results has been attempted, this report obviously fails to meet criterion (d). A graph of angle of refraction against angle of incidence *may* have shown some relationship that was not apparent from the table.

Criterion (e) – conclusion

Even although there is no clear link obvious from the results, a *directional* relationship could be observed. If the conclusion had said '. . . as the angle of incidence increases, the angle of refraction also increases. . .', it would be acceptable. Avoid the vague word 'changes'. Fails criterion (e).

Criterion (f) – evaluation

This is somewhat 'waffly' but the report does give supporting arguments for the points made – criterion (f) is met.

This report therefore meets criteria (c) and (f) but needs more work to meet criteria (b), (d) and (e). It must be improved and resubmitted.

How the count rate from a beta source changes with the thickness of aluminium shields

Aim The aim of this experiment is to find out how the radiation of a source varies with the thickness of aluminium shields.

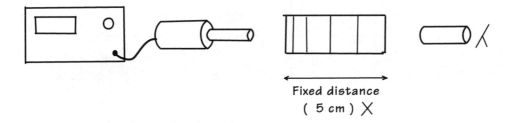

Fixed distance
(5 cm) ✗

Method The apparatus was set up as shown above, but the source not the aluminium sheets were used. The background radiation was measured and repeated three times. Then the source was placed 5 cm from the GM tube and the thickness of aluminium was varied, ✗ and was repeated to give a series of results. The count rate per minute was record-ed in each case.

Results Distance from source 5cm.

Thickness of Al (cm)	Counts per minute ✗				
	1	2	3	Average	Minus Background ✗
0	1696	1542	1626	1621.33	1577.33
0.1016	572	496	506	524.67	480.67
0.2032	108	124	130	120.67	76.67
0.3175	052	044	056	50.67	6.67
0.4572	052	052	048	50.67	6.67
0.5207	040	040	044	41.33	−3.33
Background radiation	50	40	42	44	

Conclusion From the graph, we can see that as the thickness increases the radiation emitted decreases ✗ but there is no obvious connection from the graph.

Evaluation All the readings were taken after a fixed time of 30 seconds, but it was not very accurate because you would stop a few minutes before or after 30 seconds.

There isn't much that can be done for improvement apart from the precision in time.

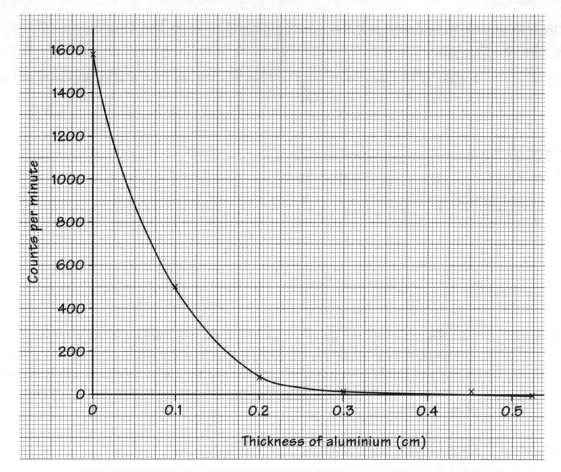

Criterion (b) – experimental procedure

1. There is neither an apparatus list nor any labels to the diagram – one or other is necessary to allow a reader to repeat the experiment successfully.
2. The fixed distance of 5 cm on the diagram should be the distance from the Geiger–Müller tube to the beta source (both unnamed) not the thickness of the aluminium shields (again unnamed).
3. There is no mention of *how* the thickness of aluminium was measured.

All three factors mean that this report fails to meet criterion (b).

Criterion (c) – recorded observations

Units have been given in the table of results, but no heading. This one often causes difficulty – count rate (heading) in counts per minute (unit). Fails criterion (c).

Criterion (d) – analysis of results

The same problem as in (c) above also occurs in the extension to the table 'minus background' (better called 'corrected count rate') and one axis of the graph: the unit has been given but the quantity has not been named. A second failure on basically the same point – fails to meet criterion (d).

Criterion (e) – conclusion

The conclusion made is not valid so the report also fails criterion (e). It is not the *radiation emitted* that decreases as the thickness increases – it is the count rate recorded at the detector that decreases.

Criterion (f) – evaluation

Surely 'a few *minutes* before or after 30 seconds' should be 'a few *seconds* before or after 30 seconds', and also some comment about the seemingly negative value of corrected count rate detected for the greatest thickness of aluminium would be expected. Fails to meet criterion (f).

Another report which fails to meet any of the criteria. In this report, most of the failings can easily be rectified. The most crucial mistake is the misunderstanding in the conclusion.

SUGGESTED ASSESSED PRACTICALS

Suggested assessed practical 1

How the instantaneous speed of an accelerating object changes with time

Apparatus: Electronic timing device (e.g. computer and interface or dedicated micro-processor)
Light gate and power supply
Trolley with single mask and trolley board
Stopclock

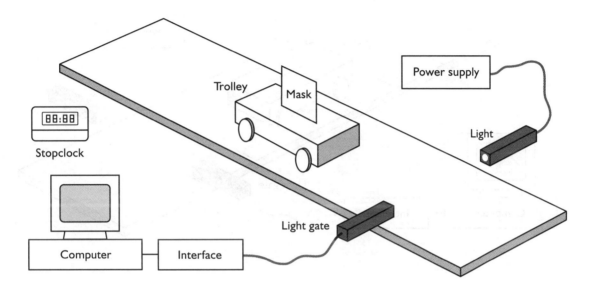

Instructions:

1. Set up the apparatus as shown in the diagram with the trolley board at a shallow angle and the light gate about one quarter of the way down the board.

2. Select a suitable program on the timing device to measure instantaneous speed and input any values necessary (e.g. width of mask).

3. Release the trolley from the top of the board and use the stopclock to measure the time it takes to reach the light gate. Record this time.

4. Read and record the instantaneous speed of the trolley at the light gate.

5. Move the light gate a short distance further down the board.

6. Repeat steps 3, 4 and 5 as necessary to obtain a series of readings.

7. Use an appropriate format to show the relationship between the instantaneous speed of an accelerating object and time.

Suggested assessed practical 2

Measuring the acceleration of a trolley released at different places on a slope

Apparatus: Electronic timing device (e.g. computer and interface or dedicated micro-processor)
Light gate and power supply
Trolley with double mask and trolley board
Metre stick

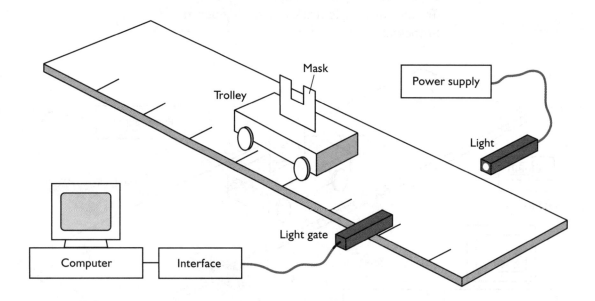

Instructions:

1. Set up the apparatus as shown in the diagram with the trolley board at a suitable angle.

2. Select a suitable program on the timing device to measure acceleration and input any values necessary (e.g. width of each section of the mask).

3. Measure and record suitable distances down the board for the light gate to be placed at and position the light gate at the first one.

4. Release the trolley from the top of the board and read and record the acceleration of the trolley at the light gate.

5. Move the light gate to the next mark down.

6. Repeat steps 4 and 5 as necessary to obtain a series of readings.

7. Use an appropriate format to show how the acceleration of an object varies as its distance of release on a slope changes.

Suggested assessed practical 3

Measuring the acceleration of a trolley at different angles of slope

Apparatus:
Electronic timing device (e.g. computer and interface or dedicated micro-processor)
Light gate and power supply
Trolley with double mask and trolley board
Large protractor

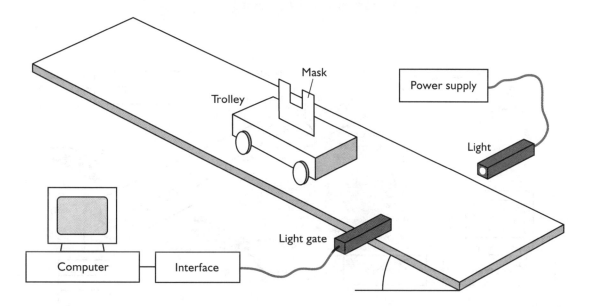

Instructions:

1. Set up the apparatus as shown in the diagram with the light gate near the bottom of the trolley board and the board at a suitable angle. Measure and record this angle.

2. Select a suitable program on the timing device to measure acceleration and input any values necessary (e.g. width of each section of the mask).

3. Release the trolley from the top of the board and read and record the acceleration of the trolley at the light gate.

4. Alter the angle of the trolley board and measure and record the new angle.

5. Repeat steps 3 and 4 as necessary to obtain a series of readings.

6. Use an appropriate format to show how the acceleration of an object varies as the angle of slope changes.

Suggested assessed practical 4

How the acceleration of an object changes as the unbalanced force is changed

Apparatus: Electronic timing device (e.g. computer and interface or dedicated micro-processor)
Light gate and power supply
Linear air track with pulley and air blower
Vehicle with double mask
10 g mass carrier and 10 g masses

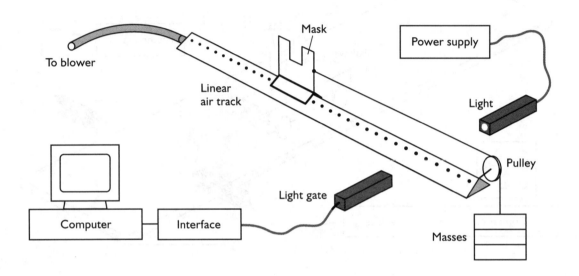

Instructions:

1. Set up the apparatus as shown in the diagram with only the 10 g mass carrier attached to the thread. Place the remaining masses on the vehicle. Make sure the linear air track is level.

2. Select a suitable program on the timing device to measure acceleration and input any values necessary (e.g. width of each section of the mask).

3. Calculate and record the weight hanging over the pulley. This is the unbalanced force.

4. Turn on the air blower and release the vehicle.

5. Read and record the acceleration of the vehicle at the light gate.

6. Transfer one 10 g mass from the vehicle to the mass carrier.

7. Repeat steps 3 to 6 as necessary to obtain a series of readings.

8. Use an appropriate format to show the relationship between the acceleration of an object and the unbalanced force applied.

Suggested assessed practical 5

How the acceleration of an object changes as its mass is changed

Apparatus: Electronic timing device (e.g. computer and interface or dedicated micro-processor)
Light gate and power supply
Linear air track with pulley and air blower
Vehicles of known masses with double mask
30 g mass on carrier

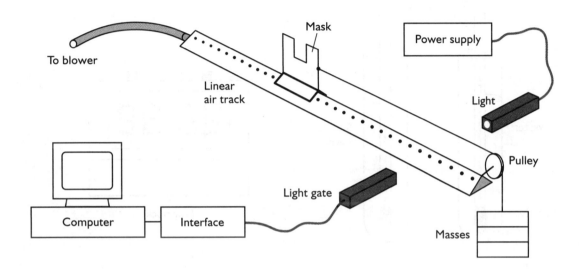

Instructions:

1. Set up the apparatus as shown in the diagram making sure the linear air track is level. Attach the 30 g mass carrier to one of the vehicles.

2. Select a suitable program on the timing device to measure acceleration and input any values necessary (e.g. width of each section of the mask).

3. Calculate and record the total mass of the vehicle and the mass carrier. This is the mass being accelerated by the unbalanced force.

4. Turn on the air blower and release the vehicle.

5. Read and record the acceleration of the vehicle at the light gate.

6. Change the mass of the vehicle by using a different one or by using a combination of vehicles joined together.

7. Repeat steps 3 to 6 as necessary to obtain a series of readings.

8. Use an appropriate format to show the relationship between the acceleration of an object and its mass.

Suggested assessed practical 6

Investigating cooling curves

Apparatus: Boiling tube containing stearic acid
Boiling tube containing water
2 thermometers
Water bath
Test tube rack
Stopclock

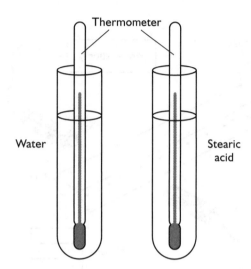

Stopclock

Instructions:

1. Heat both boiling tubes and their contents in the water bath until they are at a temperature of about 80 °C.

2. Remove both boiling tubes and place them in the test tube rack.

3. Measure and record the starting temperatures of the contents of both tubes.

4. Record the temperatures of both substances every minute for a period of about 20 minutes.

5. Use an appropriate format to show the change in the temperature of both substances with time.

Suggested assessed practical 1

How the current through a resistor changes with potential difference

Apparatus:
Resistor
Ammeter and voltmeter
Variable power supply
Connecting leads

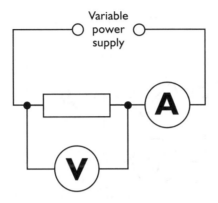

Instructions:

1. Set up the circuit shown above but do not switch on the power supply.

2. Adjust the variable power supply to give a nominal output voltage (i.e. the voltage read on the dial) of 1 V. Switch on.

3. Read and record the current through the resistor and the potential difference across it.

4. Adjust the variable power supply to increase the nominal output voltage by 1 V.

5. Repeat steps 3 and 4 until the nominal output voltage is 5 V.

6. Use an appropriate format to show the relationship between the current through a resistor and the potential difference across it.

Suggested assessed practical 2

How the resistance of a lamp changes with the current through it

Apparatus: 12 V, 24 W lamp
Ammeter and voltmeter
Variable power supply up to 12 V at 2 A
Connecting leads

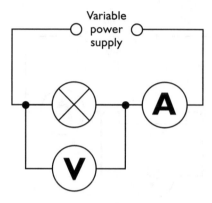

Instructions:

1. Set up the circuit shown above but do not switch on the power supply.

2. Adjust the variable power supply to give a nominal output voltage (i.e. the voltage read on the dial) of 2 V. Switch on.

3. Read and record the current through the lamp and the potential difference across it.

4. Calculate and record the resistance of the lamp.

5. Adjust the variable power supply to increase the nominal output voltage by 2 V.

6. Repeat steps 3 to 5 until the nominal output voltage is 12 V.

7. Use an appropriate format to show how the resistance of a lamp varies as the current through it changes.

Suggested assessed practical 3

The relationship between peak voltage and the quoted value of voltage

Apparatus:
Oscilloscope
a.c. power supply
a.c. voltmeter (or a multimeter on a.c. voltage range)
Resistor
Connecting leads

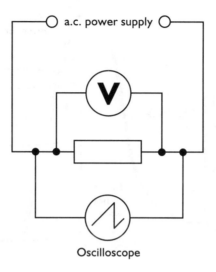

Oscilloscope

Instructions:

1. Set up the circuit shown above but do not switch on the power supply.

2. Adjust the a.c. power supply to give a nominal output voltage (i.e. the voltage read on the dial) of 1V. Switch the power supply on.

3. Adjust the oscilloscope to give a suitable trace if necessary.

4. Read and record the quoted value of voltage as shown on the voltmeter.

5. Calculate and record the peak voltage using the display shown on the oscilloscope.

6. Adjust the power supply to increase the nominal output voltage by 1V.

7. Repeat steps 3 to 6 until the nominal output voltage is 5V.

8. Use an appropriate format to show the relationship between peak voltage and the quoted value of voltage.

Suggested assessed practical 4

How the resistance of a thermistor changes as its temperature changes

Apparatus: Beaker of hot water (about 80°C)
Bead thermistor and leads
Ohmmeter
Thermometer

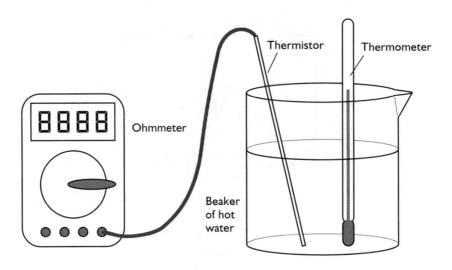

Instructions:

1. Set up the apparatus as shown in the diagram.

2. Read and record the temperature of the water.

3. Read and record the resistance of the thermistor at the same time.

4. Reduce the temperature of the water by adding cold water. Stir the water gently to achieve an even temperature.

5. Repeat steps 2 to 4 until there are six sets of results.

6. Use an appropriate format to show how the resistance of a thermistor varies as temperature changes.

Suggested assessed practical 5

The relationship between V_{out} and V_{in} for an amplifier

Apparatus: Amplifier
Signal generator
Two oscilloscopes or one dual beam oscilloscope
Connecting leads

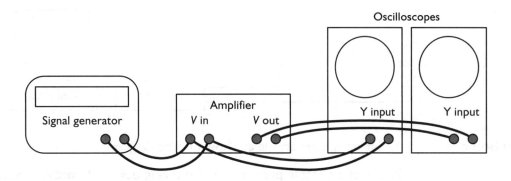

Instructions:

1. Set up the apparatus as shown in the diagram.

2. Adjust the signal generator to give a suitable low value of nominal output signal (as read on the dial) at a frequency of 1000 Hz.

3. Adjust the oscilloscope(s) to give suitable traces, if necessary.

4. Read and record both the values V_{in} and V_{out} for the amplifier.

5. Adjust the signal generator to increase its nominal output signal by a suitable amount at the same frequency.

6. Repeat steps 3 to 5 until just before the point where the output signal from the amplifier becomes distorted.

7. Use an appropriate format to show the relationship between V_{out} and V_{in} for an amplifier.

Suggested assessed practical 1

Measuring the speed of sound in air

Apparatus: Electronic timing device (e.g. computer and interface and suitable software or dedicated microprocessor)
2 sound operated switches with microphones
Means of making a sharp sound
Metre stick

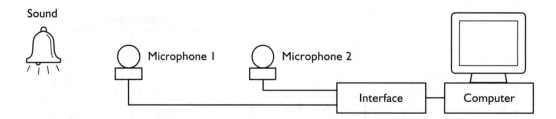

Instructions:
1. Set up the apparatus as shown in the diagram.

2. Select a suitable program on the timing device to measure a short time interval.

3. Set the two microphones 1 m apart.

4. Reset the two sound switches if necessary. Make a sharp sound to the left of microphone 1 as in the diagram.

5. Record the separation of the microphones and read and record the time interval.

6. Increase the separation of the microphones by 50 cm.

7. Repeat steps 4 to 6 until the separation of the microphones is 4 m.

8. Use an appropriate format to find the speed of sound in air.

Suggested assessed practical 2

How the angle of refraction varies with the angle of incidence

Apparatus: Semicircular glass or perspex block
Ray box and power supply
Protractor

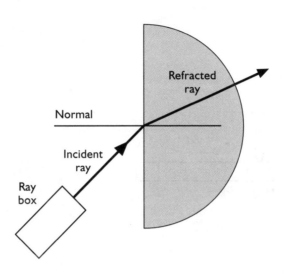

Instructions:

1. Set up the apparatus as shown in the diagram above.

2. Position the ray box to direct the incident ray to the centre of the straight edge of the block at an angle of incidence of 10°.

3. Record the angle of incidence and measure and record the angle of refraction.

4. Increase the angle of incidence by 10°.

5. Repeat steps 3 and 4 until the angle of incidence is 80°.

6. Use an appropriate format to show how the angle of refraction varies with the angle of incidence.

Suggested assessed practical 3

Measuring the focal length and calculating the power of various lenses

Apparatus: Various converging lenses
Relatively distant source of light (e.g. a window at the opposite side of the room)
Screen
Metre stick

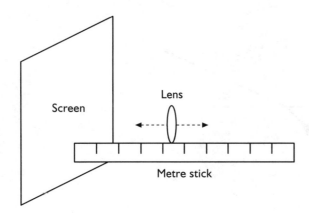
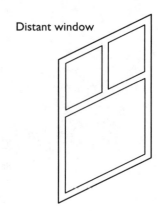

Instructions:

1. Fix a screen to the wall opposite the window in the room.

2. Choose a converging lens and move it in front of the screen until a sharp image of the distant window is seen on the screen.

3. Measure and record the focal length of the lens.

4. Calculate and record the power of the lens.

5. Repeat steps 2 to 4 for all of the lenses available.

6. Use an appropriate format to show the focal length and power of all of the lenses.

Suggested assessed practical 1

Finding the half-life of a radioactive source

Apparatus: Geiger–Müller tube
Scaler counter or ratemeter
Sealed protactinium-234 radioactive source and drip tray

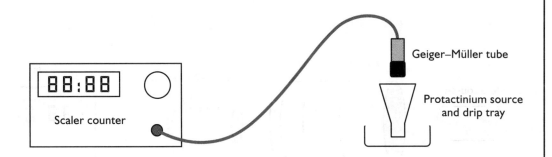

88:88

Scaler counter

Geiger–Müller tube

Protactinium source
and drip tray

Instructions:

1. Use the Geiger–Müller tube and scaler counter to measure the background count rate. Record this value.

2. Set up the apparatus shown in the diagram.

3. Measure and record values of count rate and time interval for a suitable time period.

4. Use an appropriate format to find the half-life of the radioactive source.

    Note: For safety reasons it is likely that this experiment is carried out as a teacher demonstration. Alternatively a video showing the experiment and giving suitable results may be used. Both of these methods are acceptable for this assessed practical.

Suggested assessed practical 2

Investigating how the count rate from a beta source changes with the thickness of an aluminium shield

Apparatus:
Geiger–Müller tube
Scaler counter or ratemeter
Beta emitting radioactive source
Aluminium absorbers of various thicknesses

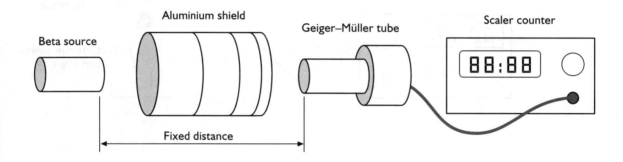

Instructions:

1. Use the Geiger–Müller tube and scaler counter to measure the background count rate. Record this value.

2. Set up the apparatus shown in the diagram with no aluminium shield initially.

3. Measure and record the value of the count rate over a suitable time interval.

4. Increase the thickness of the aluminium shield by using the absorbers either singly or in combination. Measure and record this value.

5. Repeat steps 3 and 4 until all the possible thicknesses of shield have been used.

6. Use an appropriate format to show how the count rate from a beta source changes with the thickness of an aluminium shield.

Note: For safety reasons it is likely that this experiment is carried out as a teacher demonstration. Alternatively a video showing the experiment and giving suitable results may be used. Both of these methods are acceptable for this assessed practical.

Suggested assessed practical 3

Investigating how the count rate from a gamma source changes with the thickness of a lead shield

Apparatus: Geiger–Müller tube
Scaler counter or ratemeter
Gamma emitting radioactive source
Lead absorbers of various thicknesses

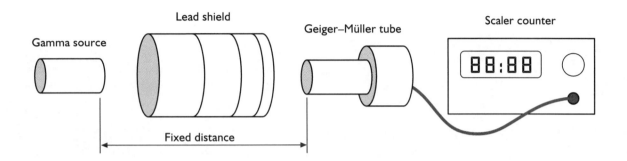

Instructions:

1. Use the Geiger–Müller tube and scaler counter to measure the background count rate. Record this value.

2. Set up the apparatus shown in the diagram with no lead shield initially.

3. Measure and record the value of the count rate over a suitable time interval.

4. Increase the thickness of the lead shield by using the absorbers either singly or in combination. Measure and record this value.

5. Repeat steps 3 and 4 until all the possible thicknesses of shield have been used.

6. Use an appropriate format to show how the count rate from a gamma source changes with the thickness of a lead shield.

   Note: For safety reasons it is likely that this experiment is carried out as a teacher demonstration. Alternatively a video showing the experiment and giving suitable results may be used. Both of these methods are acceptable for this assessed practical.

THE COURSE ASSESSMENT

Although you can gain awards for individual Units of work in the Intermediate 2 Course, it is likely that you will also go on to sit the Course assessment (the 'exam'). The details of this assessment have been given earlier ('The overall Course assessment', page 1), as have the additional skills that it tests, beyond those of the Unit tests.

In this section, there are examples of the types of questions that would be in the Course exam. The questions that follow add up to 50 marks; the final exam contains 100 marks worth of questions.

The actual question paper consists of the following types of questions.

- 20 objective questions, each worth 1 mark. These could be multiple-choice, multiple-response, true-false or matching type questions. With objective questions, there is only one correct answer.
- Questions requiring a short answer of a few words.
- Questions which are answered by doing a numerical calculation.
- Questions which need a response of up to about a paragraph.

The last three types of questions could be asked as part of a structured question built around a common part of physics – as the questions which follow show.

The questions in the final exam will cover work from all four Units in the Course. They will also test the knowledge, skills and abilities that you have gained in the Course in the following categories:

- using quantities and units correctly
- using relationships and doing calculations correctly
- describing principles correctly
- describing models (Radioactivity Unit only)
- selecting and presenting information
- processing and analysing information (including calculations)
- drawing valid conclusions and giving explanations
- describing and evaluating experimental procedures.

Although the answers have been given to the questions, along with hints on how questions would be marked, it would be useful to work through the paper without referring to the answers at first.

Sample Course assessment

1. Which of the following pairs of physical quantities are measured in the same unit?
   - (i) kinetic energy and work done.
   - (ii) heat and temperature.
   - (iii) weight and force.

   A (i) only
   B (ii) only
   C (iii) only
   D (i) and (iii) only
   E (i), (ii) and (iii)

2. Which of the following statements about the n-channel enhancement mode MOSFET is/are correct?
   - (i) It is a type of transistor.
   - (ii) It has three terminals called the drain, the gate and the source.
   - (iii) It can be used as a switch.

   A (i) only
   B (ii) only
   C (iii) only
   D (i) and (iii) only
   E (i), (ii) and (iii)

3. Which two of the following quantities used in radioactivity have units which are equivalent to the J/kg?

   absorbed dose
   activity
   equivalent dose
   radiation weighting factor

4. A model train set has a track laid out as shown.

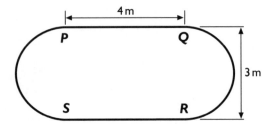

   The total length of the track is 18 m.

   (a) During one run the train took 3 minutes to make 10 complete circuits of the track at constant speed starting and finishing at point P.

   (i) Describe the motion of the train
   - (A) during section PQ of the run,
   - (B) during section SP of the run. **2**
   (ii) What distance did the train cover during this run? **1**
   (iii) What was the average speed of the train? **2**
   (iv) What was the average velocity of the train? **2**

   (b) Calculate the displacement of point R from point P on the track. **4**

   (c) The electric motor in the train is connected to the secondary winding of a 230 V mains transformer which has a turns ratio of 20 : 1.
   Calculate the output voltage of this transformer. **2**

   Total 13

5. A torch uses a lamp which is marked 3 V, 0.2 W.

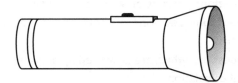

   (a) (i) Calculate the resistance of this lamp. **2**
   (ii) Draw a diagram of a circuit which could be used to check the value of this resistance by measuring the potential difference across and the current through the lamp. **3**

   (b) The original lamp in this torch is replaced by one which is marked 3 V, 0.3 W.
   What difference, if any, would this make to the operation of the torch?
   Explain your answer. **2**

   (c) The circuit diagram of the torch is shown.

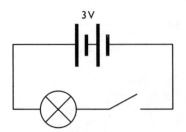

   What is the potential difference across the terminals of the *switch* when:
   - (i) the torch is off;
   - (ii) the torch is on? **2**

(d) There is a curved reflector behind the lamp. Explain, with the aid of a diagram, the function of this curved reflector.　**2**

Total 11

6. An astronomical reflecting telescope has two interchangeable eyepiece lenses. These two lenses have focal lengths of +5 mm and +20 mm respectively.

(a) Which of the following optical components must this telescope have as well as an eyepiece lens?
a mirror, an objective lens, a prism　**1**

(b) (i) Calculate the power of each of the eyepiece lenses.　**3**
(ii) By using your answer to **(b) (i)** or otherwise, explain which of the two eyepiece lenses would give the greater amount of magnification when used in the telescope.　**2**

(c) A student observes that the 20 mm lens can be used on its own as a magnifying glass.
Draw a ray diagram to show how this lens can be used in this way.　**3**

(d) The telescope is used to view a star which is 6×10^{16} m away from Earth.
(i) Calculate to the nearest year how many years after being emitted the light from this star is observed on Earth.　**3**
(ii) Suggest why the metre is not used as the unit of distance for astronomical measurements.　**1**

Total 13

7. The diagram shows part of the fission reactor at the centre of a nuclear power station.

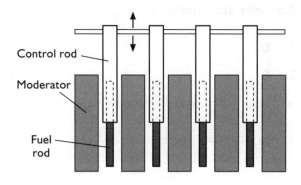

(a) What is meant by the term 'fission'?　**1**

(b) Explain the purpose of each of the three parts of a nuclear reactor shown in the diagram – the fuel rods, the moderator and the control rods.　**6**

(c) Workers in the nuclear industry have to wear radiation film badges.
Describe what such a badge is and explain how it indicates the intensity of radiation the wearer has been exposed to.　**3**

Total 10

ANSWERS TO SAMPLE COURSE ASSESSMENT

Answers and marks

Notes on marking/general comments

1. **D**   1   *heat in joules, temperature in °C*

2. **E**   1

3. absorbed dose and equivalent dose   1   *since H = D w_R and w_R has no units (No ($\frac{1}{2}$) mark)*

4. **(a)** (i) (A) constant velocity **(1)**   *because AB is a straight section of track.*
                                                                   (0) marks for 'constant speed' since this is only repeating information given in the question

           (B) changing velocity   *because the direction is always changing over section*
           OR acceleration **(1)**   2   *DA*

      (ii) distance = 10 circuits of 18 m
            = 180 m   1

      (iii) average speed = $\dfrac{\text{total distance}}{\text{total time}}$ ($\frac{1}{2}$)   *($\frac{1}{2}$) for formula*

             = $\dfrac{180}{3 \times 60}$ ($\frac{1}{2}$)   *($\frac{1}{2}$) for substitution*

             = 1 m/s **(1)**   2   *($\frac{1}{2}$) for final answer including unit, ($\frac{1}{2}$) off if unit is wrong or missing*

      (iv) ave. velocity = $\dfrac{\text{total displacement}}{\text{total time}}$ ($\frac{1}{2}$)   *($\frac{1}{2}$) for formula*

            since total displacement = 0 ($\frac{1}{2}$)   *($\frac{1}{2}$) for substitution*

            average velocity = 0 **(1)**   2   *(1) for final answer*

   (b)

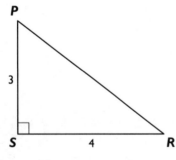

(2) marks for calculating the size of the displacement

(2) marks for calculating the direction of the displacement and stating it sensibly or showing it on a diagram

Marks would be similarly awarded if the displacement was found using a scale diagram

   PR = √(PS² + SR²) = √(9 + 16)
   = √25 = 5 m **(2)**

   tan R = $\frac{3}{4}$
   so angle R = $\tan^{-1}\frac{3}{4}$ = 37° **(2)**   4

   (c) Vp / Vs = turns ratio ($\frac{1}{2}$)   *($\frac{1}{2}$) for formula*
            so 230 / V_s = 20 ($\frac{1}{2}$)   *($\frac{1}{2}$) for substitution*
            so V_s = 230/20 = 11.5 V **(1)**   2   *(1) for final answer including unit, ($\frac{1}{2}$) off if unit is wrong or missing*

Total 13

Answers and marks

Notes on marking/general comments

5. (a) (i) $P = \dfrac{V^2}{R}$ $\left(\frac{1}{2}\right)$

$\left(\frac{1}{2}\right)$ *for formula*

$R = \dfrac{V^2}{P}$

$= \dfrac{3^2}{0.2}$ $\left(\frac{1}{2}\right)$

$\left(\frac{1}{2}\right)$ *for substitution*

$= 45\,\Omega$ **(1)**   **2**

(1) *for final answer including unit,* $\left(\frac{1}{2}\right)$ *off if unit is wrong or missing*

(ii)

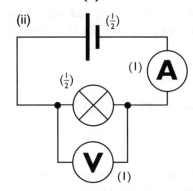

$\left(\frac{1}{2}\right)$ *for symbol,* $\left(\frac{1}{2}\right)$ *for position of each meter*
$\left(\frac{1}{2}\right)$ *for some form of power supply*
$\left(\frac{1}{2}\right)$ *for including lamp*

3

(b) Torch (lamp) would be brighter **(1)**
since the power rating is greater **(1)**   **2**

First mark would be awarded independently of second (explanation) mark, because of the way the question is worded

(c) (i) 3V **(1)**

Switch is open when torch is off

(ii) 0V **(1)**   **2**

Switch is closed when torch is on

(d)

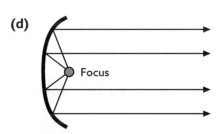

Focus

Both of the marks would be awarded for a neat, correct diagram only

The diagram is essential because of the wording of the question

(1) for bulb at focus
(1) for nearly parallel rays of light   **2**

Total 11

6. (a) a mirror   **1**

It is a reflecting telescope.

(b) (i) $P = \dfrac{1}{f}$ $\left(\frac{1}{2}\right)$

$\left(\frac{1}{2}\right)$ *for formula*

$f = 5\,\text{mm} = 5 \times 10^{-3}\,\text{m}$ $\left(\frac{1}{2}\right)$

$\left(\frac{1}{2}\right)$ *for converting to metres (only awarded once)*

$P = \dfrac{1}{5 \times 10^{-3}} = 200\,\text{D}$ **(1)**

(1) *for each final answer including unit,* $\left(\frac{1}{2}\right)$ *off if unit is wrong or missing*

$f = 20\,\text{mm} = 20 \times 10^{-3}\,\text{m}$

$P = \dfrac{1}{20 \times 10^{-3}} = 50\,\text{D}$ **(1)**   **3**

(ii) the 5 mm lens has a greater power $\left(\frac{1}{2}\right)$

Partial marking is used when the answer is incomplete

so it would refract light more $\left(\frac{1}{2}\right)$

so it would give the greater
magnification **(1)**   **2**

Answers and marks	Notes on marking/general comments

(c)

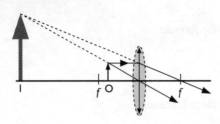

(1) for ray through centre, correctly drawn
(1) for ray parallel to axis, correctly drawn
(1) for projecting back to locate image   **3**

The position of the object is not stated with respect to the lens, so a scale diagram is not required

If the object distance was drawn greater than the focal length, (0) marks (not operating as a magnifying glass)

Each correctly drawn ray gains a mark if neatly drawn, using a ruler. Marks would be deducted for poorly drawn rays (light does travel in straight lines, after all)

(d) (i)
$$t = \frac{s}{v} \ \left(\tfrac{1}{2}\right)$$

$$= \frac{6 \times 10^{16}}{3 \times 10^{8}} \ \left(\tfrac{1}{2}\right) + \left(\tfrac{1}{2}\right)$$

$$= 2 \times 10^{8} \text{ s} \ \left(\tfrac{1}{2}\right)$$

$$\text{so } t = \frac{2 \times 10^{8}}{60 \times 60 \times 24 \times 365}$$

$$= 6.34 \text{ years} \ \left(\tfrac{1}{2}\right)$$

time = 6 years to nearest year $\left(\tfrac{1}{2}\right)$ **3**

$\left(\tfrac{1}{2}\right)$ *for formula*

$\left(\tfrac{1}{2}\right)$ *for data (3 x 10⁸ m/s), $\left(\tfrac{1}{2}\right)$ for correct substitution*

$\left(\tfrac{1}{2}\right)$ *for final answer – no unit mark since it was given in the question but $\left(\tfrac{1}{2}\right)$ would be deducted if the wrong unit was given*

(ii) Astronomical distances are so great that the metre is too small to be convenient.   **1**

Total 13

Any reasonable way of wording this answer would be awarded the mark

7. (a) Fission is the breaking up of the nucleus of an atom into smaller nuclei.   **1**

(b) The fuel rod contains the radioactive material (usually uranium dioxide) **(1)**
which produces neutrons by fission. **(1)**

It is not necessary to name the fuel used, since it is not asked for

The moderator slows down the fast-moving neutrons (by successive collisions) **(1)**
allowing them to become thermal neutrons **or** so that they can cause further fission. **(1)**

How the moderator does its job (by successive collisions) is not asked for, so is not needed

The control rods regulate the number of (thermal) neutrons **(1)**
and so control the rate of the reaction. **(1)**   **6**

Again, how the control rods do their job of controlling is not required

(c) A radiation film badge contains a photographic film **(1)**
behind different absorbers of radiation. **(1)**
The amount of fogging at each part indicates the intensity of radiation.   **(1) 3**

Total 10

(3) marks for a question like this indicates the amount of detail to be included in the answer

ANSWERS TO SELECTED QUESTIONS
ANSWERS TO MECHANICS AND HEAT QUESTIONS

Kinematics

5.   distance = 200 m
     time = 25 s

     average speed = $\dfrac{\text{total distance travelled}}{\text{total time taken}}$

     $\bar{v} = \dfrac{s}{t}$ $\bar{v} = \dfrac{200}{25} = 8$

     $\bar{v} = \underline{8 \text{ m/s}}$

6.   distance = 72 km = 72 000 m
     time = $1\frac{1}{4}$ hours = 75 minutes
           = 75 × 60 s = 4500 s

     average speed = $\dfrac{\text{total distance travelled}}{\text{total time taken}}$

     $\bar{v} = \dfrac{s}{t}$ $\bar{v} = \dfrac{72\,000}{4500} = 16$

     $\bar{v} = \underline{16 \text{ m/s}}$

7.   time = 1 minute = 60 s
     speed = 340 m/s

     average speed = $\dfrac{\text{total distance travelled}}{\text{total time taken}}$

     $\bar{v} = \dfrac{s}{t}$

     so distance = speed × time
     s = 340 × 60 = 20 400
     s = $\underline{20\,400 \text{ m}}$ (= $\underline{20.4 \text{ km}}$)

10. (a) distance = 30 km
          time = $\frac{1}{2}$ hour

          average speed = $\dfrac{\text{total distance travelled}}{\text{total time taken}}$

          $\bar{v} = \dfrac{s}{t}$ $\bar{v} = \dfrac{30}{\frac{1}{2}} = 60$

          $\bar{v} = \underline{60 \text{ km/h}}$

     (b) The instantaneous speed will be 0 when
          the bus is stationary or picking up passen-
          gers. It will be higher than the average at
          times on the motorway and will have vari-
          ous values up to the speed limit in town
          traffic.

     (c) The instantaneous speed constantly
          changes, depending on the conditions. The
          average speed is calculated over the whole
          journey, and takes account of varying
          instantaneous speeds.

13.

Scalar quantities	Vector quantities
speed	velocity
energy	acceleration due
work done	to gravity
temperature	weight
heat	gravitational field
time	strength
mass	

16. (a) speed
     (b) velocity
     (c) speed
     (d) speed
     (e) velocity

19. The acceleration of a car which is travelling at a
     steady speed along a straight level road is zero,
     since the rate of change of speed is zero.

20. (a) A negative sign associated with an acceler-
          ation value shows that the object is slow-
          ing down.

     (b) Another term used for a negative acceler-
          ation is a deceleration.

21. Performance figures such as '0 to 26 m/s in 8.2
     seconds' refer to the maximum acceleration of
     a car.

23. acceleration = $\dfrac{\text{change in velocity}}{\text{time for change}}$

          $= \dfrac{\text{final velocity} - \text{initial velocity}}{\text{time for change}}$

     $a = \dfrac{v - u}{t}$

     so $v - u = at$ $\underline{\text{so } v = u + at}$

24. change in velocity $(v - u) = 10$ m/s
     time = 5 s

     acceleration = $\dfrac{\text{change in velocity}}{\text{time for change}}$

     so $a = \dfrac{10}{5} = 2$

     $a = \underline{2 \text{ m/s}^2}$

25. initial velocity $u = 0$
final velocity $v = 26$ m/s
time $t = 8$ s

$$\text{acceleration} = \frac{\text{change in velocity}}{\text{time for change}}$$

$$= \frac{\text{final velocity} - \text{initial velocity}}{\text{time for change}}$$

so $a = \dfrac{26 - 0}{8} = 3.25$

$a = \underline{3.25 \text{ m/s}^2}$

26. $a = 0.6$ m/s^2
$u = 2$ m/s
$t = 30$ s
$v = u + at$ so $v = 2 + (0.6 \times 30)$
$\qquad\qquad\qquad\quad = 2 + 18 = 20$
$v = \underline{20 \text{ m/s}}$

27. $u = 0$ ('from rest')
$v = 6$ m/s
$t = 3$ s
$a = \dfrac{v - u}{t}$
so $a = \dfrac{6 - 0}{3} = 2$
$a = \underline{2 \text{ m/s}^2}$

28. $a = 0.1$ m/s^2
$u = 1$ m/s
$v = 5$ m/s
$v = u + at$ so $5 = 1 + (0.1 \times t)$
so $0.1 \times t = 4$ so $t = \underline{40 \text{ s}}$

29. $a = -2$ m/s^2 (deceleration, so negative)
$v = 0$ ('to rest')
$t = 10$ s
$v = u + at$ so $0 = u + ((-2) \times 10)$
so $u = \underline{20 \text{ m/s}}$

30.

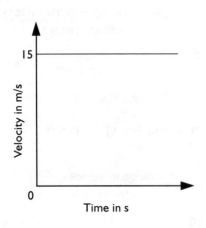

31.

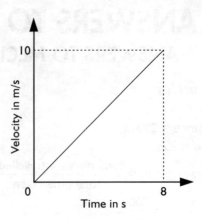

32.

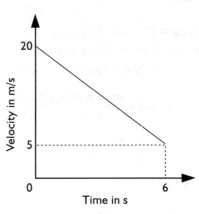

33. (i) Velocity–time graph of an object travelling at a steady velocity of 20 m/s. Since the velocity of the object does not change, its acceleration is zero.

(ii) Velocity–time graph of an object increasing velocity from 0 to 20 m/s in 4 s.
$u = 0$
$v = 20$ m/s
$t = 4$ s
$a = \dfrac{v - u}{t}$
so $a = \dfrac{20 - 0}{4} = 5$
$a = \underline{5 \text{ m/s}^2}$

(iii) Velocity–time graph of an object increasing velocity uniformly from 10 m/s to 20 m/s in 2 s.
$u = 10$ m/s
$v = 20$ m/s
$t = 2$ s
$a = \dfrac{v - u}{t}$
so $a = \dfrac{20 - 10}{4} = \dfrac{10}{2} = 5$
$a = \underline{5 \text{ m/s}^2}$

(iv) Velocity–time graph of an object decreasing velocity from 20 m/s to rest in 5 s.

$u = 20$ m/s

$v = 0$

$t = 5$ s

$a = \dfrac{v - u}{t}$

so $a = \dfrac{0 - 20}{5} = -\dfrac{20}{5} = -4$

$a = \underline{-4 \text{ m/s}^2}$

This is a negative acceleration or a deceleration of 4 m/s^2

35. (a) During section AB the car increases its velocity uniformly from 0 to 15 m/s in 3 s.
During section BC the car travels at a steady velocity of 15 m/s for 5 s.
During section CD the car decreases its velocity uniformly from 15 m/s to 0 in 6 s.

(b) The maximum velocity reached by the car is 15 m/s.

(c) The car accelerates during section AB of the graph.

$u = 0$

$v = 15$ m/s

$t = 3$ s

$a = \dfrac{v - u}{t}$

so $a = \dfrac{15 - 0}{3} = 5$

$a = \underline{5 \text{ m/s}^2}$

(d) The car decelerates during section CD of the graph.

$u = 15$ m/s

$v = 0$

$t = (14 - 8) = 6$ s

$a = \dfrac{v - u}{t}$

so $a = \dfrac{0 - 15}{6} = -2.5$

$a = \underline{-2.5 \text{ m/s}^2}$

(e) The displacement of the car is equal to the area under the graph.

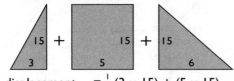

displacement $= \frac{1}{2}(3 \times 15) + (5 \times 15)$
$\qquad\qquad + \frac{1}{2}(6 \times 15)$
$\qquad\quad = 22.5 + 75 + 45 = 142.5$

displacement $= \underline{142.5 \text{ m}}$

36. (a)

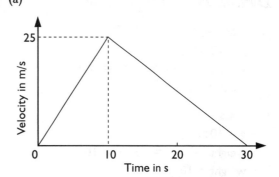

(b) $u = 0$

$v = 25$ m/s

$t = 10$ s

$a = \dfrac{v - u}{t}$

so $a = \dfrac{25 - 0}{10} = 2.5$

$a = \underline{2.5 \text{ m/s}^2}$

(c) $u = 25$ m/s

$v = 0$

$t = (30 - 10) = 20$ s

$a = \dfrac{v - u}{t}$

so $a = \dfrac{0 - 25}{20} = -1.25$

$a = \underline{-1.25 \text{ m/s}^2}$

(d) Distance stations are apart is equal to the area under the graph.

distance $= \frac{1}{2}(10 \times 25) + \frac{1}{2}(20 \times 25)$
$\qquad\quad = 125 + 250$

distance $= \underline{375 \text{ m}}$

37. (a)

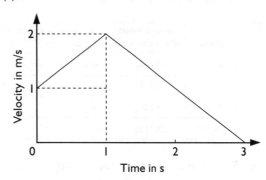

(b) Total length of runway is equal to the area under the graph.

length $= (1 \times 1) + \frac{1}{2}(1 \times 1) + \frac{1}{2}(2 \times 2)$
$\qquad\quad = 1 + \frac{1}{2} + 2 = 3.5$

length $= \underline{3.5 \text{ m}}$

Dynamics

16. mass = 50 kg

$g = 10 \, \text{N/kg}$

weight = $m\,g$ = 50 × 10 = 500

weight = $\underline{500\,\text{N}}$

17. mass = 1 kg

$g = 10 \, \text{N/kg}$

weight = $m\,g$ = 1 × 10 = 10

weight = $\underline{10\,\text{N}}$

18. weight = 100 000 N

$g = 10 \, \text{N/kg}$

weight = $m\,g$

so mass = $\dfrac{\text{weight}}{g} = \dfrac{100\,000}{10} = 10\,000$

mass = $\underline{10\,000\,\text{kg}}$

19. (a) When the block is taken to the Moon its mass stays the same and its weight decreases.

(b) When the block is taken to Jupiter its mass stays the same and its weight increases.

(c) When the block is taken into space far away from any planets its mass stays the same and its weight becomes zero.

20. weight on Earth = 25 N

gravitational field strength on Earth = 10 N/kg

weight = mass × gravitational field strength

so mass = $\dfrac{25}{10} = 2.5$ mass = 2.5 kg

gravitational field strength on Moon = 1.6 N/kg

so weight on Moon = 2.5 × 1.6 = $\underline{4\,\text{N}}$

21.

Planet	Gravitational field strength	Weight of spaceman of mass 120 kg
Venus	9 N/kg	$m\,g$ = 120 × 9 = $\underline{1080\,\text{N}}$
Earth	10 N/kg	$m\,g$ = 120 × 10 = $\underline{1200\,\text{N}}$
Mars	4 N/kg	$m\,g$ = 120 × 4 = $\underline{480\,\text{N}}$
Jupiter	26 N/kg	$m\,g$ = 120 × 26 = $\underline{3120\,\text{N}}$

28. Since the book is not moving, its speed is constant (zero) and so the forces acting on it are balanced.

29. (a) The forces acting on a car which is travelling along a straight level road at a constant speed are balanced.

(b) The car engine needs to be on to maintain the steady speed because the engine has to supply a force to balance the force of friction trying to slow it down.

30. You continually have to pedal a bicycle to move along a straight level road at a constant speed because a force has to be supplied to balance the force of friction trying to slow the bicycle down.

39. When a second person jumps on to a moving skateboard, the mass increases and so it will decelerate since the pushing force remains the same.

40. (a) If the driver of a car which is travelling at a constant speed along a straight, level road supplies an unbalanced force to it, the car will accelerate.

(b) The pedal in a car that is used to increase the unbalanced force on the car is called the accelerator.

41. If a balloonist throws a sandbag overboard from a hot air balloon which is falling at a steady speed in the air, the mass of the balloon is decreased. As a result of this decrease in mass the weight also decreases so the upward air resistance is now greater than the downward weight causing an upward unbalanced force. Both of these effects cause the balloon to have an upward acceleration.

43. (a)   (b)

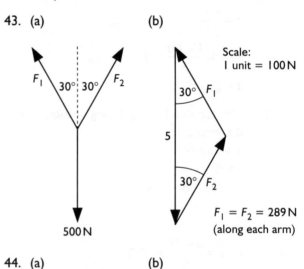

44. (a)   (b)

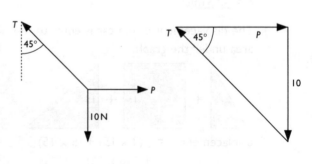

46.

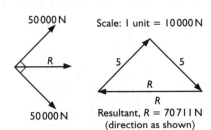

Scale: 1 unit = 10 000 N

Resultant, $R = 70711 N$
(direction as shown)

49. mass = 0.75 kg
unbalanced force = 3 N

$$\text{acceleration} = \frac{\text{unbalanced force}}{\text{mass}} \qquad a = \frac{F}{m}$$

so $a = \frac{3}{0.75} = 4$

$a = \underline{4 \text{ m/s}^2}$

50. acceleration = 5 m/s²
mass = 2 kg
unbalanced force = mass x acceleration
so $F = m\,a = 2 \times 5 = 10$
$F = \underline{10 \text{ N}}$

51. unbalanced force = −1500 N (negative since it
causes a deceleration)
deceleration = 2 m/s² (so $a = -2$ m/s²)

$$\text{acceleration} = \frac{\text{unbalanced force}}{\text{mass}}$$

so mass = $\dfrac{\text{unbalanced force}}{\text{acceleration}}$

so $m = \dfrac{F}{a} = \dfrac{-1500}{-2} = 750$

$m = \underline{750 \text{ kg}}$

52. mass = 100 kg

(a) weight = $m \times g = 100 \times 10 = \underline{1000 \text{ N}}$

(b) acceleration = 0.1 m/s² downwards
unbalanced force = mass x acceleration
so $F = m\,a = 100 \times 0.1 = \underline{10 \text{ N downwards}}$

(c)

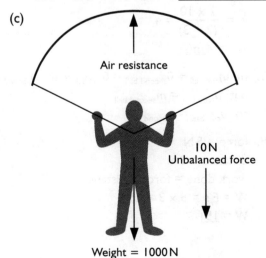

Air resistance

10 N
Unbalanced force

Weight = 1000 N

unbalanced force = weight − air resistance
so air resistance = weight − unbalanced
force
= 1000 − 10 = 990
air resistance = $\underline{990 \text{ N}}$

53. thrust = 2000 N
mass = 5000 kg

$$\text{acceleration} = \frac{\text{force}}{\text{mass}} = \frac{2000}{5000} = 0.4$$

acceleration = $\underline{0.4 \text{ m/s}^2}$

54. (a) A rocket motor does not need to be kept
on all the time while the rocket is moving
far away from any planets because there are
no forces to change its motion. There is no
wind or air resistance since space is a vacuum and there is no gravitational pull from
any planet. Since there are no forces acting
on the rocket, it will continue to move in a
straight line at a steady speed. This is an
example of Newton's First Law.

(b) If the rocket motor was fired there would
be an unbalanced force acting on the rocket and so it would accelerate − either
change its speed or change its direction.

60. (a) Since the horizontal velocity of a projectile
is constant, the horizontal velocity of the
parcel just before it reaches the ground is
$\underline{50 \text{ m/s}}$.

(b) Horizontally:
$v = 50$ m/s
$t = 4$ s
$v = \dfrac{s}{t}$

so $s = v\,t = 50 \times 4 = 200$
$s = \underline{200 \text{ m}}$

(c) Vertically:
$u = 0$ (the parcel starts out with horizontal
motion only)
$v = ?$ (this is the vertical velocity of the parcel just before it reaches the ground)
$a = 10$ m/s² (the acceleration due to
gravity)
$t = 4$ s (the time is common to horizontal
and vertical motion)
$v = u + a\,t$
so $v = 0 + (10 \times 4) = 40$
$v = \underline{40 \text{ m/s}}$

(d) To find the height of the helicopter when the parcel was dropped (which is the same as the vertical distance travelled by the parcel), sketch the graph of the vertical velocity of the parcel against time.

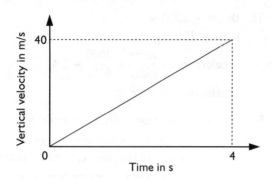

The height of the helicopter is given by the area under the graph.
height = $\frac{1}{2}$ (4 x 40)
so height = $\underline{80\,m}$

Momentum and energy

3. (a) How a car moves forward along a road

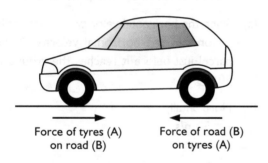

Force of tyres (A)
on road (B)

Force of road (B)
on tyres (A)

(b) How a chair can support a person sitting on it

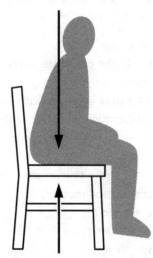

Force of chair (A)
on person (B)

(c) How a rocket moves forward in space.

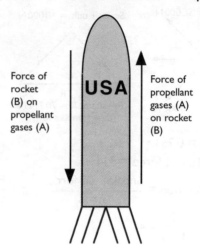

Force of rocket (B) on propellant gases (A)

USA

Force of propellant gases (A) on rocket (B)

5. The 'Newton Pairs' acting on the boat are as follows.

Upthrust: The force of the water (A) on the boat (B) and the force of the boat (B) on the water (A).

Weight: The force of the Earth (A) on the boat (B) and the force of the boat (B) on the Earth (A).

Force on sail: The force of the air (A) on the sail (B) and the force of the sail (B) on the air (A).

Resistive force: The force of friction of the water (A) on the boat (B) and the force of the boat (B) on the water (A).

11. $m_1\,u_1 = (m_1 + m_2)\,v_2$
$$v_2 = \frac{1000 \times 0.51}{(1000 + 700)} = 0.3$$
$v_2 = \underline{0.3\,m/s}$

12. $m_1\,u_1 = (m_1 + m_2)\,v_2$
$$v_2 = \frac{2 \times 10}{(2 + 3)} = 4$$
$v_2 = \underline{4\,m/s}$

13. $(m\,u)_{cue\ ball} + 0_{second\ ball} = 0_{cue\ ball} + (m\,v)_{second\ ball}$
but $m_{cue\ ball} = m_{second\ ball}$
so $u_{cue\ ball} = v_{second\ ball}$

18. force = 5 N
distance = 3 m
work done = force x distance
$W = F\,s = 5 \times 3 = 15$
$W = \underline{15\,J}$

19. energy (work done) = 1000 J

distance = 20 m

work done = force x distance

so force = $\dfrac{\text{work done}}{\text{distance}} = \dfrac{1000}{20} = 50$

force = <u>50 N</u>

20. work done = 200 J

force = 40 N

work done = force x distance

so distance = $\dfrac{\text{work done}}{\text{force}} = \dfrac{200}{40} = 5$

force = <u>5 m</u>

23. work done = 360 J

time = 1 minute = 60 s

power = $\dfrac{\text{work done}}{\text{time}} = \dfrac{360}{60} = 6$

power = <u>6 W</u>

24. mass = 60 kg

time = 10 s

height = 5 m

g = 10 m/s² (not stated explicitly)

(a) weight = $m\,g$ = 60 x 10 = <u>600 N</u>

(b) work done = $F\,s$ = weight x height
   = 600 x 5 = <u>3000 J</u>

(c) power = $\dfrac{\text{work done}}{\text{time}} = \dfrac{3000}{10} = $ <u>300 W</u>

25. mass = 0.3 kg

g = 10 m/s² (not stated explicitly)

velocity = 20 cm/s = 0.2 m/s

power = force x velocity = weight x velocity

= $m\,g\,v$ = 0.3 x 10 x 0.2 = 0.6

power = <u>0.6 W</u>

31. mass = 50 kg

height = 1.5 m

g = 10 m/s² (not stated explicitly)

$E_p = m\,g\,h$ = 50 x 10 x 1.5 = 750

E_p = <u>750 J</u>

36. mass = 0.75 kg

speed = 2 m/s

$E_k = \frac{1}{2}\,m\,v^2 = \frac{1}{2}$ x 0.75 x 2² = 1.5

E_k = <u>1.5 J</u>

37. Car: Lorry:

mass = 800 kg   mass = 3 tonnes = 3000 kg

speed = 26 m/s   speed = 13 m/s

$E_k = \frac{1}{2}\,m\,v^2$ $E_k = \frac{1}{2}\,m\,v^2$

$= \frac{1}{2}$ x 800 x 26² $= \frac{1}{2}$ x 3000 x 13²

E_k = 270 400 J E_k = 253 500 J

The car has the greater amount of kinetic energy.

40. P_{in} = 3000 W

P_{out} = 2700 W

percentage efficiency = $\dfrac{P_{out}}{P_{in}}$ x 100

$= \dfrac{2700}{3000}$ x 100

= <u>90%</u>

41. E_{in} = 10 J

$E_{out} = m\,g\,h$ = 0.1 x 10 x 2 = 2 J

percentage efficiency = $\dfrac{E_{out}}{E_{in}}$ x 100

$= \dfrac{2}{10}$ x 100

= <u>20%</u>

Heat

12. $E_h = c\,m\,\Delta T$

so E_h = 902 x 1 x 10 = <u>9020 J</u>

13. $E_h = c\,m\,\Delta T$

so E_h = 4180 x 2 x (90 − 20)

= 4180 x 2 x 70

so E_h = <u>585 200 J</u> (585.2 kJ)

14. $E_h = c\,m\,\Delta T$

so 720 000 = 2400 x 5 x ΔT

so $\Delta T = \dfrac{720\ 000}{2400\ \times\ 5} = $ <u>60 °C</u>

18.

23. (a) Another word which means the same as 'fusion' is melting.

(b) Another term which is often used for 'vaporisation' is evaporation.

27. $E_h = m\,l$

so E_h = 0.5 x 2.26 x 10⁶

so E_h = <u>1.13 x 10⁶ J</u>

28. $E_h = m\,l$

so E_h = 0.5 x 3.34 x 10⁵

so E_h = <u>1.67 x 10⁵ J</u>

29. total heat needed =

heat to increase temperature of ice from −18 °C to 0 °C ($c\,m\,\Delta T$)$_{ice}$

+ heat to melt ice to water at 0 °C ($m\,l$)$_{fusion}$

+ heat to increase temperature of water from 0 °C to 100 °C ($c\,m\,\Delta T$)$_{water}$

+ heat to evaporate water to steam at 100 °C

$(m\ l)_{vaporisation}$

= 2100 x 1 x (0 – (–18)) [= 37 800 J]

   + 1 x 3.34 x 10^5 [= 334 000 J]

   + 4180 x 1 x (100 – 0) [= 418 000 J]

   + 1 x 2.26 x 10^6 [= 2 260 000 J]

total = <u>3 049 800 J</u>

31. heat needed: $E_h = c\ m\ \Delta T$

          E_h = 4180 x 1.5 x (100 – 20)

          E_h = 4180 x 1.5 x 80

          E_h = 501 600 J

this heat is supplied by electricity

so $P \times t$ = 501 600

2200 x t = 501 600

so $t = \dfrac{501\ 600}{2200}$ = <u>228 s</u> (3 min 48 s).

32. E_k of car $= \frac{1}{2}\ m\ v^2 = \frac{1}{2}$ x 1000 x 4^2 = 8000 J

all of this energy is converted into heat

so $E_h = c\ m\ \Delta T$ = 8000

so 500 x 0.5 x ΔT = 8000

so $\Delta T = \dfrac{8000}{500 \times 0.5}$

so ΔT = <u>32 °C</u>

33. mass = 3 kg

speed = 2000 m/s

so kinetic energy $E_k = \frac{1}{2}\ m\ v^2 = \frac{1}{2}$ x 3 x 2000^2

E_k = 6 x 10^6 J

specific heat capacity of iron, c_{iron}, = 440 J/kg °C

if 10% of the kinetic energy is converted into heat, then

E_k x 10% = $c\ m\ \Delta T$

so 6 x 10^5 = 440 x 3 x ΔT

so $\Delta T = \dfrac{6 \times 10^5}{440 \times 3}$ = <u>455 °C</u>

34. If air resistance can be ignored, all of the potential energy of an object (E_p = $m\ g\ h$) is transformed into kinetic energy (E_k $= \frac{1}{2}\ m\ v^2$) when it falls.

so $E_p = E_k$

so $m\ g\ h = \frac{1}{2}\ m\ v^2$

but, since the mass is the same, $g\ h = \frac{1}{2}v^2$

so $v^2 = 2\ g\ h$

so <u>$v = \sqrt{(2\ g\ h)}$</u>

35. height = 80 m

g = 10 m/s^2 (not stated explicitly)

(a) E_k at bottom = E_p at top

so $v = \sqrt{(2\ g\ h)} = \sqrt{(2 \times 10 \times 80)} = \sqrt{1600}$

v = <u>40 m/s</u>

(b) If air resistance cannot be ignored the speed of the stone as it enters the water will be less than 40 m/s since not all of the potential energy will be transformed into kinetic energy.

(c) The main form of energy, other than kinetic energy, that is produced in practice is **heat**.

36. (a) A spacecraft has kinetic energy because of its movement.

(b) This kinetic energy is changed into heat when the spacecraft re-enters the Earth's atmosphere from space.

(c) The friction as the spacecraft moves quickly through the Earth's atmosphere causes this energy transformation to take place.

37. mass = 70 000 kg

velocity in orbit = 8000 m/s

velocity at touchdown = 100 m/s

kinetic energy $= \frac{1}{2}\ m\ v^2$

(a) kinetic energy of the orbiter while in orbit

$= \frac{1}{2}$ x 70 000 x 8000^2

so E_k = <u>2.24 x 10^{12} J</u>

(b) kinetic energy of the orbiter at touchdown

$= \frac{1}{2}$ x 70 000 x 100^2

so E_k = <u>3.5 x 10^8 J</u>

(c) The 'lost' kinetic energy of the orbiter has been converted into heat, because the orbiter has been slowed down by friction as it entered the Earth's atmosphere.

(d) kinetic energy of the orbiter at touchdown

= 3.5 x 10^8 J (from part (b))

average force needed to stop orbiter = 175 kN = 175 000 N

work done in stopping orbiter (force x distance) = kinetic energy at touchdown

so 175 000 x distance = 3.5 x 10^8

so distance $= \dfrac{3.5 \times 10^8}{175\ 000}$ = 2000

distance = <u>2000 m</u> (= <u>2 km</u>)

ANSWERS TO ELECTRICITY AND ELECTRONICS QUESTIONS

Circuits

9. charge = 180 C
   time = 1 minute = 60 s

   $$I = \frac{Q}{t} = \frac{180}{60} = \underline{3A}$$

31.

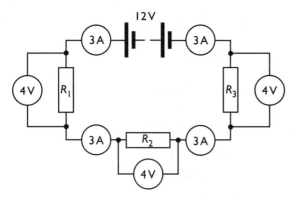

35.

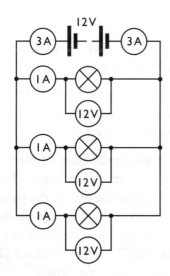

40. voltage = 9 V
    resistance = 180 Ω

    $$\text{resistance} = \frac{\text{voltage}}{\text{current}} \quad \text{so} \quad 180 = \frac{9}{\text{current}}$$

    so   current = $\frac{9}{180}$ = $\underline{0.05\ A}$ (= $\underline{50\ mA}$)

41. resistance = 1 kΩ = 1000 Ω
    current = 10 mA = 10 × 10⁻³ A
    potential difference = current × resistance
    $V = I R = 10 \times 10^{-3} \times 1000 = \underline{10\,V}$

43. $R_T = R_1 + R_2$
    $= 82 + 18$
    $R_T = \underline{100\,\Omega}$

44. $R_T = R_1 + R_2 + R_3$
    $= 2.2 + 4.7 + 5.6$
    $R_T = \underline{12.5\,\Omega}$

46. $\dfrac{1}{R_T} = \dfrac{1}{R_1} + \dfrac{1}{R_2}$

    $= \dfrac{1}{12} + \dfrac{1}{15}$

    so $R_T = \underline{6.7\,\Omega}$

47. $\dfrac{1}{R_T} = \dfrac{1}{R_1} + \dfrac{1}{R_2} + \dfrac{1}{R_3}$

    $= \dfrac{1}{10} + \dfrac{1}{12} + \dfrac{1}{15}$

    so $R_T = \underline{4\,\Omega}$

51. $V_1 = V_s \times \dfrac{R_1}{R_1 + R_2}$

    $= 9 \times \dfrac{220}{220 + 180}$

    so $V_1 = \underline{4.95\,V}$

52. If V_1 is half of V_s then $V_1 = V_2 = V_s/2$
    so $R_1 = R_2$
    The two resistors have equal values.

Electrical energy

7. current = 0.25 A
   voltage = 3 V
   power = voltage × current
   so $P = 3 \times 0.25 = \underline{0.75\,W}$

9. voltage = mains voltage = 230 V
   power = 460 W
   power = voltage × current
   so   460 = 230 × current

   so current = $\dfrac{460}{230}$ = $\underline{2A}$

11. resistance = 10 kΩ = 10 × 10³ Ω
    current = 5 mA = 5 × 10⁻³ A
    power (rate of energy transformation) = $I^2 R$
    so $P = (5 \times 10^{-3})^2 \times 10 \times 10^3 = \underline{0.25\,W}$

12. power = 5 W
    voltage = 12 V

    $P = \dfrac{V^2}{R} \qquad 5 = \dfrac{12^2}{R} \qquad R = \dfrac{144}{5}$

    $R = \underline{28.8\,\Omega}$

14. (a) resistance = 92 Ω

voltage = 230 V

$$I = \frac{V}{R} = \frac{230}{92}$$

$I = \underline{2.5\,A}$

(b) Use a <u>3A fuse</u> (since the current is less than 3 A)

Electromagnetism

15. V_p = 230V (**mains** transformer)

V_s = 11.5V

n_p = 2000

$$\frac{V_p}{V_s} = \frac{n_p}{n_s}$$

so $n_s = \frac{V_s}{V_p} \times n_p = \frac{11.5}{230} \times 2000 = \underline{100\ turns}$

18. (a) step-down

(b) turns ratio = voltage ratio

$$= \frac{V_p}{V_s} = \frac{230}{115} = 2{:}1$$

(c) $P_{in} = P_{out}$ so $I_p\,V_p = I_s\,V_s$

$$I_p = \frac{12 \times 115}{230} = \underline{6\,A}$$

21.

Resistance = 0.2 Ω/km

Transmission lines

Length = 25 km

Resistance per kilometre = 0.2 Ω/km

total length = 2 x 25 km (2 transmission lines)

total length = 50 km

so total resistance = resistance per kilometre x length = 0.2 x 50 = 10 Ω

current = 20 A

power loss = $I^2 R = 20^2 \times 10 = 4000$

power loss = <u>4000 W</u> = <u>(4 kW)</u>

Electronic components

10. (a) LED

(b) seven-segment display

(c) loudspeaker

(d) solenoid

(e) relay

17. current = 12.5 mA = 12.5×10^{-3} A

potential difference across resistor = 9 − 2

= 7V

$$R = \frac{7}{12.5 \times 10^{-3}} = 560$$

resistance of series resistor = <u>560 Ω</u>

26. (a) LDR

(b) microphone

(c) solar cell

(d) thermistor

(e) thermocouple

35. when R = 1000 Ω:

$$I = \frac{V}{R} = \frac{5}{1000} = 5$$

$I = \underline{5\,mA}$

when R = 400 Ω:

$$I = \frac{V}{R} = \frac{5}{400} = 12.5$$

$I = \underline{12.5\,mA}$

36. (a) when I = 10 mA:

$$R = \frac{V}{I} = \frac{2}{10 \times 10^{-3}} = 200$$

$R = \underline{200\ \Omega}$

when I = 25 mA:

$$R = \frac{V}{I} = \frac{2}{25 \times 10^{-3}} = 80$$

$R = \underline{80\ \Omega}$

(b) The light intensity increases because the resistance decreases.

43. (a) This circuit lights the LED (4) when the temperature of the thermistor (1) falls below a certain value.

(b) When the temperature of the thermistor falls, its resistance increases. This causes the voltage at the base of the transistor (3) to rise. When this voltage reaches about 0.7 V, the transistor conducts and the LED emits light. The variable resistor (2) adjusts the sensitivity of the circuit.

44. (a) This circuit lights the LED (4) when the intensity of light falling on the LDR (1) rises above a certain value.

(b) When the intensity of light falling on the LDR rises, its resistance decreases. This causes the voltage at the base of the transistor (3) to rise. When this voltage reaches about 0.7 V, the transistor conducts and the LED emits light. The variable resistor (2) adjusts the sensitivity of the circuit.

45. (a) This circuit lights the LED (4) when the intensity of light falling on the LDR (1) falls below a certain value.

(b) When the intensity of light falling on the LDR falls, its resistance increases. This causes the voltage at the base of the transistor (3) to rise. When this voltage reaches about 0.7 V, the transistor conducts and the LED emits light. The variable resistor (2) adjusts the sensitivity of the circuit.

52. input voltage = 10 mV = 10×10^{-3} V
output voltage = 0.5 V

$$\text{voltage gain} = \frac{\text{output voltage}}{\text{input voltage}} = \frac{0.5}{10 \times 10^{-3}} = \underline{50}$$

ANSWERS TO WAVES AND OPTICS QUESTIONS

Waves

13. distance = 10 m
    time = 4 s
    $$v = \frac{s}{t} = \frac{10}{4} = 2.5$$
    $v = \underline{2.5\,\text{m/s}}$

14. time = 5 s
    wave speed = 3 m/s
    $$v = \frac{s}{t} \quad \text{so } s = v\,t = 3 \times 5 = 15$$
    $s = \underline{15\,\text{m}}$

15. speed of sound = 340 m/s
    time = 5 s
    $$v = \frac{s}{t}$$
    so $s = v\,t = 340 \times 5 = 1700$
    $s = \underline{1700\,\text{m}} \ (= \underline{1.7\,\text{km}})$

16. time = 1.2 s
    speed = 3×10^8 m/s
    $$v = \frac{s}{t}$$
    so $s = v\,t = 3 \times 10^8 \times 1.2 = 3.6 \times 10^8$ m
    $s = \underline{3.6 \times 10^8\,\text{m}}$

17. $s = 150\,000\,000$ km $= 1.5 \times 10^{11}$ m
    $v = 3 \times 10^8$ m/s
    $$v = \frac{s}{t} \quad \text{so } t = \frac{s}{v} = \frac{1.5 \times 10^{11}}{3 \times 10^8} = 500$$
    $t = \underline{500\,\text{s}} \ (\underline{8\ \text{minutes}\ 20\,\text{s}})$

19. time (for sound to travel) = 3 s
    speed of sound in air = 340 m/s
    $$v = \frac{s}{t}$$
    so $s = v\,t = 340 \times 3 = 1020$
    $s = \underline{1020\ \text{m}}$

20. speed of sound in air = 340 m/s
    distance = 1.36 km = 1360 m
    $$v = \frac{s}{t} \quad \text{so } t = \frac{s}{v} = \frac{1360}{340} = 4$$
    $t = \underline{4\,\text{s}}$

34.

Quiet sound,
low frequency

Quiet sound,
high frequency

Loud sound,
low frequency

Loud sound,
high frequency

36. frequency = 262 Hz
    wave speed = 340 m/s
    $$v = f\lambda \quad \text{so } \lambda = \frac{v}{f} = \frac{340}{262} = 1.297\,709\,9$$
    $\lambda = \underline{1.30\,\text{m}}$

37. wavelength = 50 cm = 0.5 m
    speed = 25 cm/s = 0.25 m/s
    $$v = f\lambda \quad \text{so } f = \frac{v}{\lambda} = \frac{0.25}{0.5} = 0.5$$
    $f = \underline{0.5\ \text{times a second}} \ (\underline{0.5\ \text{Hz}})$

40. $v = 3 \times 10^8$ m/s
    $\lambda = 660$ nm $= 660 \times 10^{-9}$ m
    $$v = f\lambda \quad \text{so } f = \frac{v}{\lambda} = \frac{3 \times 10^8}{660 \times 10^{-9}} = 4.55 \times 10^{14}$$
    $f = \underline{4.55 \times 10^{14}\,\text{Hz}}$

41. frequency = 909 kHz = 909×10^3 Hz
    speed = 3×10^8 m/s
    $$v = f\lambda \quad \text{so } \lambda = \frac{v}{f} = \frac{3 \times 10^8}{909 \times 10^3} = 330$$
    $\lambda = \underline{330\ \text{m}}$

42. wavelength = 194 m
    speed = 3×10^8 m/s
    $$v = f\lambda \quad \text{so } f = \frac{v}{\lambda} = \frac{3 \times 10^8}{194} = 1.546 \times 10^6$$
    $f = \underline{1546\ \text{kHz}}$

Reflection

12. visible light; microwaves; TV waves; radio waves; sound; infrared signals

21. signal transmission in telecommunications; telephones

Refraction

10. A refracting telescope uses lenses to produce images by refraction of light rays.

24. focal length = 25 cm = 0.25 m
    $$\text{power} = \frac{1}{\text{focal length}} = \frac{1}{0.25} = 4$$
    $P = \underline{+4\,\text{D}}$

25. power = −20 D
    $$\text{focal length} = \frac{1}{\text{power}} = \frac{1}{-20} = -0.05$$
    $f = \underline{-0.05\,\text{m}} \ (\underline{-5\,\text{cm}})$

Dosimetry

16. absorbed dose $D = 40\,\mu Gy$
     radiation weighting factor for alpha particles
     $w_R = 20$
     equivalent dose $H = D\,w_R$
     $H = 40 \times 20$
     $H = \underline{800\,\mu Sv}$

17. equivalent dose $H = D\,w_R$
     $H = (200 \times 3) + (50 \times 1)$
     $H = \underline{650\,\mu Sv}$

Half-life and safety

5.   The half-life of the source is 15 days, so 60 days
     is equal to 4 half-lives.
     original activity　　　　　$= 1600$ kBq
     activity after 1 half-life　$= \frac{1}{2} \times 1600 = 800$ kBq
     activity after 2 half-lives $= \frac{1}{2} \times 800\;\; = 400$ kBq
     activity after 3 half-lives $= \frac{1}{2} \times 400\;\; = 200$ kBq
     activity after 4 half-lives $= \frac{1}{2} \times 200\;\; = \underline{100\ kBq}$

6.   (a)

Week	1	2	3	4	5	6	7	8	9
Recorded activity in counts/minute	140.0	82.0	50.0	33.0	23.5	18.0	15.5	14.0	13.0
Corrected activity in counts/minute	128.0	70.0	38.0	21.0	11.5	6.0	3.5	2.0	1.0

(b)

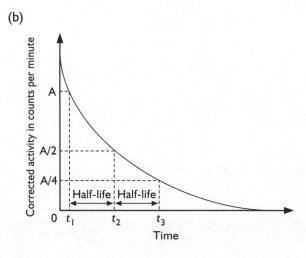

(c)　about 8 days

INDEX

Major page references for index entries are in bold.

ACKNOWLEDGEMENTS

The publishers would like to thank the following for permission to reproduce photographs on the pages listed:

Alan Thomas: pp. 8, 12 (foot), 20, 22, 36, 38, 40, 41, 43, 45, 46, 47, 48, 67, 81

Empics/ Tony Marshall: p. 10

NASA/Science Photo Library: p. 12 (top right)

Kent Wood/Science Photo Library: p. 60

BSIP Laurent/H. Americain/Science Photo Library: p.79

Philippe Plaitly/Eurelios/Science Photo Library: p. 80

David Parker/Science Photo Library: p. 83 (top left)

Simon Fraser/Science Photo Library: p. 83 (right)

Novosti/Science Photo Library: p. 85